The RFID Roadmap: The Next Steps for Europe

Gerd Wolfram · Birgit Gampl · Peter Gabriel
Editors

The RFID Roadmap: The Next Steps for Europe

Springer

Dr. Gerd Wolfram
MGI METRO Group Information
Technology GmbH
Metro-Straße 12
40235 Düsseldorf
Germany
gerd.wolfram@mgi.de

Dr. Birgit Gampl
MGI METRO Group Information
Technology GmbH
Metro-Straße 12
40235 Düsseldorf
Germany
birgit.gampl@mgi.de

Dipl. Inf. Peter Gabriel
VDI/VDE Innovation + Technik GmbH
Steinplatz 1
10623 Berlin
Germany
gabriel@vdivde-it.de

ISBN 978-3-540-71018-9

e-ISBN 978-3-540-71019-6

DOI 10.1007/978-3-540-71019-6

© 2008 Springer-Verlag Berlin Heidelberg

This work is subject to copyright. All rights are reserved, whether the whole or part of the material is concerned, specifically the rights of translation, reprinting, reuse of illustrations, recitation, broadcasting, reproduction on microfilm or in any other way, and storage in data banks. Duplication of this publication or parts thereof is permitted only under the provisions of the German Copyright Law of September 9, 1965, in its current version, and permission for use must always be obtained from Springer. Violations are liable to prosecution under the German Copyright Law.

The use of general descriptive names, registered names, trademarks, etc. in this publication does not imply, even in the absence of a specific statement, that such names are exempt from the relevant protective laws and regulations and therefore free for general use.

Coverdesign: WMX Design GmbH, Heidelberg, according to a draft by André Zeich, VDI/VDE-IT

Printed on acid-free paper

9 8 7 6 5 4 3 2 1

springer.com

Foreword

There are few examples of new technologies that have early on been identified to possess a significant economic potential across a wide range of industries and at the same time triggered as much attention from consumer organisations and policy makers around the world as Radio Frequency Identification (RFID). Coming in different shapes, applications and use cases, RFID devices are now rapidly moving from research labs to mass deployments, hinting at the technology's potential to become an engine for growth and jobs, and to help turning Europe into a highly competitive and dynamic knowledge-based economy – the goal set in the "Lisbon Strategy".

Given the successful research and the rapid uptake of the technology as well as the public debate about its implications and potential, the European Commission must ensure that RFID can be used within a consistent framework across Europe to avoid legal, ethical and economic discrepancies.

RFID-related issues and concerns are plentiful and diverse. Thus, over the past two years and a half, Viviane Reding, Commissioner for the Information Society and Media, has launched and led an international debate on Radio Frequency Identification (RFID) by guiding thinking, providing forums for the exchange of ideas, and influencing public policy making.

The role of the European Commission is to drive and foster a coherent European approach that ensures common standards, harmonised legislation and compatible guidelines. The Commission will continue to analyse the rationale and options to create a policy framework for RFID, taking into account the ongoing open discussions with relevant stakeholders – consumers' organisations, market actors, national and European authorities and relevant international stakeholders.

As part of its ongoing effort to take all stakeholders' opinions into account, the European Commission has invited a network of key industry players to develop their recommendations for the establishment of a European framework for RFID. Technology providers and RFID users have responded and launched the initiative "Coordinating European Efforts for Promoting the European RFID Value Chain" (CE RFID).

The EU-funded project CE RFID has provided valuable insights into how to combine forces to enhance the conditions of competition for RFID technology, as well as to boost the political environment of RFID at European level. Establishing a framework and reinforcing the communication between key players are indeed essential requirements for a competitive industry and for a secure, safe and privacy-preserving deployment of RFID in Europe.

After two years of efforts, the CE RFID project has now drawn to a close and decided to publicise its most significant results. This book includes a number of recommendations for improving the benefits and reducing the problems likely to stem from advances in RFID technology. Some of these recommendations are directed at engineers, especially those involved in designing and developing RFID systems. Some of them – in particular those in chapter 5 in relation to the Regulatory Framework (i.e. privacy and security, health and environment, Intellectual Property Rights, and RFID governance) – are aimed at policy makers who need to be aware of and pro-active in RFID developments. Other recommendations are for commercial organisations which deal with RFID implementations as part of their everyday business and see a clear need for and advantage in exchanging experiences and good practices.

This report is welcomed by the European Commission as it is perceived as an important milestone and a valuable contribution to the wider debate which it has initiated and stimulated over the past few months and which is still going on.

Gérald Santucci, Ph. D.
Head of Unit
"Networked Enterprise and Radio Frequency Identification" (RFID)
Information Society and Media Directorate-General
European Commission

Preface

This book is a summary of the results elaborated within the European coordination action project CE RFID (Coordinating European Efforts for Promoting the European RFID Value Chain). CE RFID was funded within the 6th Framework Programme for Research, Technological Development and Demonstration of the European Commission and was intended to assist the European Commission in the development of a European RFID policy. The target of CE RFID was to support sustainable improvement of market conditions for RFID technology, and to strengthen its development in Europe from an industry point of view while preserving fundamental values like the respect and protection of privacy, health and environment. The book is directed towards decision makers from politics, economics, science, and administration, and provides helpful guidance on different subjects in the field of RFID. This was achieved by:

- Defining an RFID roadmap for future technology and application development
- Developing recommendations for the harmonisation of RFID radio frequencies and standards
- Examination of the legal framework with regards to the deployment of RFID in Europe including an analysis of privacy concerns
- Analysis of existing RFID implementation and application guidelines
- Analysis of specific research programmes and projects at a European level

From these objectives, the project was divided into five main RFID-related topics: roadmap, R&D research policy, standardisation, guidelines, and legislation. For each topic, detailed extended final reports have been delivered which can be downloaded from the CE RFID website (www.rfid-in-action.eu).

The following 12 companies formed the core set of CE RFID:

- MGI METRO Group Information Technology GmbH (coordinator)
- AIDA Centre S.L.
- Deutsche Post World Net
- EADS Deutschland GmbH

- FEIG Electronic GmbH
- NXP Semiconductors Austria GmbH
- Pleon GmbH
- RF-iT-Solutions GmbH
- Siemens AG
- Tyco Fire & Security ADT
- UPM Raflatac Oy
- VDI/VDE Innovation + Technik GmbH

CE RFID was established as an industry consortium designed to assess the challenges which prevent RFID development from an industry point of view. Some of the most important European companies involved in RFID were part of this consortium. During this project, they looked beyond their own company's needs and together with RFID experts from various fields developed recommendations of how to support both RFID users and technology suppliers in Europe. By doing this, a roadmap was created as to how to keep and further extend Europe's position as an innovative key player in the field of RFID technology.

The project lasted from April 2006 until September 2008. Several workshops and a review process for each topic took place to allow a feedback process and to develop consensus between the different experts. The approximately 200 participants of the workshops and the document reviews represented the relevant stakeholders in the RFID area. This helped us to achieve valuable results, and allowed us to give clear recommendations regarding which actions need to be taken by different groups to allow RFID technology to further develop. All final reports are available in the internet: www.rfid-in-action.eu.

On behalf of all project partners the editors would like to thank all participants of the workshops and the reviewers of the reports for discussing the different topics with us, and for having given us valuable input by reading the respective documents and for their time and engagement. Without their efforts it would not have been possible to integrate so many different stakeholder views.

All of the results elaborated within the CE RFID project would not have been achieved without the support of the Information Society and Media Directorate-General of the European Commission and we are grateful for this continuous support and interest in our project and the respective results. Many thanks to our project officer Cristina Martinez Gonzalez and the reviewers Dr Zoe Kardasiadou (Centre of International and European Economic Law), Dr Humberto Moran (Open Source Innovation Ltd) and John Kayser (evercom AG) for their constant support.

Neither this book nor the overall project results would have been possible without the very productive cooperation of all partners within this project. A deep engagement above and beyond the expected level of commitment, and the extensive specialist competencies possessed by members have allowed a very supportive and sustainable social network to be built, which has helped us to achieve excellent and profound project results. We would also like to offer many thanks to Matthias Robeck (MGI METRO Group Information Technology GmbH) who expertly led the teams with care and attention throughout the project and to Lysann Müller (VDI/VDE Innovation + Technik GmbH) for her diligent advice on all EU Framework Programme matters. Without their continuous support, our accomplishments would not have been possible.

Furthermore we appreciate the support of Dr Jens Strüker from the University of Freiburg, who contributed to the different topics of this project, and in addition for reading the draft of this book and giving very valuable comments. We are much obliged to Cristina Martinez Gonzalez (European Commission) and Dr Zoe Kardasiadou (Centre of International and European Economic Law) for their detailed and fruitful last review of this book. Many thanks also to Wolfram Groß (VDI/VDE Innovation + Technik GmbH) for his valuable comments on the technology roadmap. Last but not least, we would like to give thanks to Chris Hankinson (Nottingham Trent University), who supported the editors as an English native speaker student throughout the elaboration of this book, for proof reading and for his continuous and engaged assistance.

Contributors

This book has been compiled and edited by:

Dr Gerd Wolfram (MGI)
Dr Birgit Gampl (MGI)
Peter Gabriel (VDI/VDE-IT)

The different chapters of the book have been written by the following authors:

1. Introduction
   Dr Birgit Gampl (MGI), Peter Gabriel (VDI/VDE-IT), Dr Gerd Wolfram (MGI)

2. Framework for the Classification of RFID Applications and Stakeholders
   Dr Birgit Gampl (MGI), Dr Gerd Wolfram (MGI)

3. Standards
   Eldor Walk (FEIG), Alexander Gauby (RF-iT), Frank Neubauer (EADS)

4. Implementation and Application Guidelines
   Dr Birgit Gampl (MGI), Sebastian Lange (Pleon), Chris Hankinson (NTU)

5. Regulatory Framework
   Véronique Corduant (DPWN), Andreas Kruse (DPWN), Camino Mortera-Martinez (DPWN), Sebastian Lange (Pleon)

6. Technological Research Needs
   Gerhard Metz (Siemens), Peter Gabriel (VDI/VDE-IT)

7. Technological Research Needs
   Dr Klaus Pavlik (NXP)

8. Conclusion
   Dr Birgit Gampl (MGI), Dr Gerd Wolfram (MGI)

DPWN	Deutsche Post World Net
EADS	EADS Deutschland GmbH
FEIG	FEIG Electronic GmbH
MGI	MGI METRO Group Information Technology GmbH
NTU	Nottingham Business School, Nottingham Trent University
NXP	NXP Semiconductors Austria GmbH
Pleon	Pleon GmbH
RF-iT	RF-iT Solutions GmbH
Siemens	Siemens AG
VDI/VDE-IT	VDI/VDE Innovation + Technik GmbH

Table of Contents

1	**Introduction**..		1
	1.1	The Technology Diffusion of RFID and Specific Challenges.......	4
	1.2	Outline of the Book	6
2	**Framework for the Classification of RFID Applications and Stakeholders**..		9
	2.1	The RFID Reference Model	9
	2.2	The RFID Stakeholder Model	12
3	**Standards**...		15
	3.1	Standardisation Organisations and Processes	16
		3.1.1 Basic Rule Setting Organisations....................................	16
		3.1.2 Standard Development Organisations.............................	17
		3.1.3 User and Industry Organisations.....................................	18
		3.1.4 Business Models of Standardisation Organisations	18
	3.2	Radio Spectrum Framework...	20
	3.3	Interoperability of Standards	26
		3.3.1 Product Codes	27
		3.3.2 The Internet of Things..	28
		3.3.3 Data Exchange and the Object Name Service (ONS)	29
	3.4	Analysis of the Need for Application Specific Standards............	31
		3.4.1 Logistical Tracking and Tracing of Goods	32
		3.4.2 Production, Monitoring and Maintenance of Goods and Processes...	34
		3.4.3 Product Safety, Quality and Information of Goods and Processes	36
		3.4.4 Access Control Systems, Personal Tracking, Rental Systems...	38

		3.4.5	General Assessment of Current RFID Application Standards	39
		3.4.6	General Recommendations on RFID Application Standards	40
	3.5	Need for Standards for RFID Sensor Tags		40
	3.6	Privacy and Security Standards		41
		3.6.1	Privacy	41
		3.6.2	Security	42
		3.6.3	Data Security Measures in Air Interface Standards	43
		3.6.4	Recommendations on Privacy and Data Security	44

4 Implementation and Application Guidelines ... 47

	4.1	Requirements of Guidelines		48
		4.1.1	The RFID Reference Model	48
		4.1.2	The RFID Implementation Checklist	48
	4.2	Analysis of Existing Guidelines		51
		4.2.1	Method	51
		4.2.2	Initial Categorisation	53
		4.2.3	Definition of RFID Guidelines	54
		4.2.4	Process of Analysis	55
		4.2.5	List of Guidelines Analysed	56
	4.3	Quantitative Analysis of Guidelines		64
		4.3.1	Formal Categories	64
		4.3.2	Addressees of Guidelines	66
		4.3.3	Consideration of Stakeholders	70
	4.4	Establishing Guidelines Using the RFID Implementation Checklist		72
	4.5	Conclusions		77
		4.5.1	Relevance of Existing RFID Guidelines	77
		4.5.2	The RFID Implementation Checklist – next Steps	78

5 Regulatory Framework ... 81

	5.1	Privacy		81
		5.1.1	Legal Framework	82
		5.1.2	Data Protection Principles and the Definition of Personal Data	84
		5.1.3	RFID and Data Protection Legislation: a Case Specific Approach	88
		5.1.4	Conclusions	96
	5.2	Health and Environmental Effects		100
		5.2.1	Health Effects	100
		5.2.2	Environmental Effects	101

		5.3	Radio Spectrum 103
			5.3.1 EC Legislation and other Policy Texts............................ 103
			5.3.2 Analysis 106
			5.3.3 Conclusion 107
		5.4	The Intellectual Property Rights Framework............................... 108
			5.4.1 Policy Approaches 108
			5.4.2 Industry approaches 111
			5.4.3 Open Source Approach: OpenPCD................................. 113
			5.4.4 Conclusions... 113
		5.5	RFID Governance.. 115
			5.5.1 Observation of Current Debate on Internet Governance 115
			5.5.2 Legal Framework and Approaches to RFID Governance 117
			5.5.3 Conclusions... 119

6 Technological Research Needs.. 121
6.1 General Technology Challenges.. 121
6.2 Technology Requirements 123
6.2.1 Tags... 123
6.2.2 Readers.. 125
6.2.3 System Technology.. 126
6.3 RFID Technology Roadmap.. 127
6.3.1 Packaging 129
6.3.2 Chip Design 129
6.3.3 Energy Aspects 130
6.3.4 RF Technology... 131
6.3.5 Manufacturing.. 132
6.3.6 Systems 133
6.3.7 Readers.. 134
6.3.8 Non-Silicon Technologies... 135
6.3.9 Bi-stable Displays 136
6.3.10 Sensors 136
6.3.11 Cryptography 137
6.3.12 ICT Architectures.. 138
6.3.13 Environmental Aspects 138

7 R&D Environment.. 141
7.1 Outline and Approach... 141
7.1.1 Assessment Criteria of R&D Support Programmes 142
7.1.2 Methodology used for the Analysis 143
7.1.3 Programmes and Countries Considered 144

	7.2	Analysis of National Programmes	145
		7.2.1 Germany	145
		7.2.2 France	148
		7.2.3 UK	151
		7.2.4 The Netherlands	151
		7.2.5 Italy	153
		7.2.6 Spain	153
		7.2.7 Austria	154
		7.2.8 Finland	156
		7.2.9 New Member States	158
	7.3	Transnational Programmes with National Funding	159
		7.3.1 NORDITE	159
		7.3.2 EUREKA	159
	7.4	Transnational Programmes with Joint National and EU Funding	161
	7.5	European Programmes	161
	7.6	R&D Programmes & the RFID Reference Model	164
	7.7	Conclusions of RFID R&D Funding Programme Assessment	165
		7.7.1 Thematic Focus of Funded Programmes	165
		7.7.2 Funding Structures	166
	7.8	Recommendation for a Future European R&D Policy	168
8	**Conclusion: The Next Steps for Europe**		**175**
	8.1	The Fields of Activities	175
	8.2	The Stakeholder Perspective	179
References			**187**
Index			**197**

List of Tables

Table 1.1	Adoption stages	5
Table 3.1	Logistical tracking and tracing of goods overview	32
Table 3.2	Production, monitoring and maintenance of goods and processes overview	34
Table 3.3	Product safety, quality and information of goods and processes overview	36
Table 3.4	Access control systems, personal tracking, rental systems overview	38
Table 3.5	SWOT analysis of current RFID application standards	39
Table 3.6	Data security measures in interface standards	44
Table 4.1	RFID Implementation Checklist	50
Table 4.2	Example of code book	52
Table 4.3	List and description of RFID guidelines analysed	57
Table 4.4	Overview of authors	64
Table 4.5	Year of publication	65
Table 4.6	Maximum fulfilment of requirements per addressee	69
Table 4.7	Stakeholders and frequency of being considered	71
Table 4.8	SWOT analysis of RFID implementation guidelines	78
Table 5.1	RFID-related EU privacy legislation	83
Table 5.2	RFID-related EU radio spectrum legislation	104
Table 6.1	Tag classes of the EPC Roadmap	128
Table 7.1	Overview of national programmes	145
Table 7.2	Overview of NORDITE programme	159
Table 7.3	Overview of the EURKEA project	161
Table 7.4	FP5/FP6 programmes	163

xvii

List of Figures

Fig. 1.1	RFID Diffusion Model	4
Fig. 2.1	RFID Reference Model	10
Fig. 2.2	RFID Stakeholder Model	13
Fig. 3.1	RFID frequency bands	21
Fig. 3.2	Proposal for high performance RFID and SRD applications	26
Fig. 3.3	Basic types of attack on RFID systems (source: BSI 2005)	42
Fig. 4.1	Process for analysing existing guidelines	55
Fig. 4.2	Process diagram for analysis of requirements of addressees	56
Fig. 4.3	Number of pages per guideline	65
Fig. 4.4	Frequency of addressees	66
Fig. 4.5	Number of addressees addressed by guidelines	67
Fig. 4.6	Fulfilment of requirements for all addressees	68
Fig. 4.7	Analysis of requirements – according to addressees	69
Fig. 4.8	Addressees, stakeholders and possible interactions	70
Fig. 4.9	Frequency of stakeholders being considered	71
Fig. 6.1	Packaging design timeline	129
Fig. 6.2	Chip design timeline	130
Fig. 6.3	Energy aspects timeline	130
Fig. 6.4	RF Technology development timeline	131
Fig. 6.5	Manufacturing development timeline	132
Fig. 6.6	Systems development timeline	133
Fig. 6.7	Non silicon technology development timeline	134
Fig. 6.8	SAW/Polymer technology development timeline	135
Fig. 6.9	Bi-stable display development timeline	136
Fig. 6.10	Sensor tags development timeline	137
Fig. 6.11	Cryptography development timeline	137
Fig. 6.12	ICT architecture development timeline	138
Fig. 6.13	Environmental research timeline	139
Fig. 7.1	Overview of German RFID projects	146
Fig. 7.2	Overview of French RFID projects	149

Fig. 7.3	Overview of Austrian RFID projects	155
Fig. 7.4	Overview of Finnish RFID projects	157
Fig. 7.5	Overview of the EUREKA project	160
Fig. 7.6	Research domains related to RFID	162
Fig. 7.7	Overview of application areas in European programmes	162
Fig. 7.8	Overview of FP6 programmes	163

Abbreviations

ADS	Afilias Discovery Services
AFNOR	French Association of Standardisation
AIAG	Automotive Industry Action Group
AWID	Applied Wireless Identification Group
BMWi	German Federal Ministry of Economics
BSI	British Standardisation Institute
BSI	German Federal Bureau of Information Security
CCTV	Closed Circuit Television
CE	European Conformity
CEE	Central and Eastern European Countries
CE RFID	Coordinating European Efforts for Promoting the European RFID Value Chain
CEN	European Committee for Standardisation
CENELEC	European Committee for Electrotechnical Standardisation
CERP	Cluster of European RFID Projects
CII	Computer Implemented Inventions
DG INFSO	Information Society and Media Directorate General
DIN	German Standardisation Institute
DNS	Domain Name System
DWD	German Weather Service
EAN	European Article Number
ECJ	European Court of Justice
EFTA	European Free Trade Area
EMC	Electromagnetic Compatibility
EPC	Electronic Product Code
EPCIS	Electronic Product Code Information Service
EPLA	European Patent Litigation Agreement
EPO	European Patent Office
EPoSS	European Technology Platform on Smart Systems Integration
ERM	Electromagnetic Compatibility and Radio Spectrum Matters
ERP	Effective Radiated Power
ESDS	Extensible Supply Chain Discovery Service
ETP	European Technology Platform
ETSI	European Telecommunications Standardisation Institute
FP5/6/7	Framework Programme 5/6/7
FRAND	Fair, Reasonable and Non-Discriminatory Terms
FSA	Fluidic Self Assembly

GHz	Gigahertz
GLN	Global Location Number
GRAI	Global Returnable Asset Identifier
GSM	Global System for Mobile Communication
GTIN	Global Trade Item Number
HF	High Frequency
HR	Human Resources
IATA	International Air Travel Association
IC	Integrated Circuit
ICANN	Internet Corporation for Assigned Names and Numbers
ICAO	International Civil Aviation Organisation
ICT	Information and Communication Technology
IEEE	Institute of Electrical and Electronics Engineers
IoT	Internet of Things
IP	Internet Protocol
IPR	Intellectual Property Rights
ISM	Industrial, Scientific, Medical (Frequency Bands)
ISO	International Organisation for Standardisation
IT	Information Technology
ITF	Interrogator Talks First
ITU	International Telecommunications Union
ITU-R	International Telecommunications Union – Radiocommunication Sector
LAN	Local Area Network
LCD	Liquid Crystal Display
LF	Low Frequency
MEMS	Micro Electronic Mechanical System
MHz	Megahertz
MRP	Machine Readable Passport
NGO	Non Governmental Organisation
NTIA	National Telecommunications and Information Administration
OECD	Organisation for Economic Cooperation and Development
ONS	Object Name Service
PCD	Proximity Coupling Devices
PET	Privacy Enhancing Technology
PISA	Privacy Incorporated Software Agent
QAM	Quadrature Amplitude Modulation
R&D	Research and Development
RAND	Reasonable and Non-Discriminatory Terms
RF	Radio Frequency
RFID	Radio Frequency Identification
ROI	Return On Investment
SAE	Society of Automotive Engineers
SAW	Surface Acoustic Wave
SD-Card	Secure Digital Card
SDO	Standardisation Development Organisation
SME	Small to Medium Sized Enterprise
SNR	Signal to Noise Ratio
SRD	Short Range Devices
SSCC	Serial Shipping Container Code
SWOT	Strengths, Weaknesses, Opportunities, Threats
TRIPS	Trade Related Aspects of Intellectual Property
UID	Unique Identifier
UHF	Ultra High Frequency
ULD	Unit Load Device

URL	Universal Resource Locator
USPTO	United States Patent and Trademark Office
UWB	Ultra Wide Band
VCD	Vicinity Coupling Device
VDE	Association of German Electrical Engineers
VDI	Association of German Engineers
VICC	Vicinity Card
VPN	Virtual Private Network
WEEE	Waste Electrical and Electronic Equipment Directive
WGIG	Working Group on Internet Governance
WLAN	Wireless Local Area Network
WTO	World Trade Organisation

Chapter 1
Introduction

The auto identification technology RFID (Radio Frequency Identification) was first developed in the 1940's, when it was used to distinguish between friend and foe aircraft in the Second World War. Afterwards, it became a civilian technology introduced in niche applications such as animal tracking or car toll systems. In the late 1990's advancements in technology, such as the miniaturisation of integrated circuits and the etching of antennas onto flexible substrates (i.e. plastics) opened up a great number of new application fields; from contactless smart cards to electronic tickets, and even sensor tags which can monitor the temperature of fresh foods or medication, therefore allowing quality control.

Around the millennium there was hype about RFID technology, which promised a revolution in logistics. This promise has clearly proven to be an exaggerated one. However, radio frequency identification is becoming a cornerstone of modern ICT systems with great opportunities and challenges, not only for European users, but also for European companies who supply RFID technology. Revenue from radio frequency identification (RFID) technology should reach $ 1.2 billion this year worldwide, up nearly 31 % from $ 917.3 million in 2007 (Eschinger 2008a). Europe is currently below growth expectations made by industry experts. It is predicted that although the uptake of RFID will increase in 2008 and 2009, Europe will move relatively slowly with regards to high tag volume deployments. It is predicted however, that the number of innovative RFID solutions will continue to increase (Eschinger 2008a). The amount of high level RFID trials and the release of Generation 2 standards mean that European companies are beginning to evaluate the business case for RFID introduction. RFID can also provide many benefits for European citizens, such as in the field of healthcare or concerning product safety issues.

Europe as an RFID Technology User

From an economic point of view, one should distinguish between Europe's role as a technology user and as a technology provider. To begin with the user side, European companies and authorities are among the early adopters of RFID. Applications can be found in many private sectors, like automotive, retail, and logistics, as well as in the public sector. From a short-term perspective, the technology offers its users new means to increase productivity by having a better and timelier control of the flow of goods and by providing real time information about economic processes. In the long term it will create new products and services such as self-steering packages that move autonomously through an intermodal transport chain (OECD 2008).

It is a difficult and complex task to calculate the economic effects of RFID; therefore currently there are only a few estimations regarding the economic impacts of RFID from a user's perspective. A study on behalf of the German Ministry of Economics stated that by 2015, up to 8 % of the German GDP (Gross Domestic Product) would be affected by RFID. Productivity effects caused by RFID in German retail were estimated at € 8.6 billion by 2010. The corresponding figures for the German logistics and automotive sectors were € 4.3 billion and € 2.4 billion respectively (Bovenschulte et al. 2007). A study by the University of Austin estimated that the retail sector will gain a world-wide economic benefit of $ 68.55 billion by 2011 (Barua et al. 2006), whereas the health benefits are valued at around $ 34.67 billion in 2011. As preliminary as these numbers are, they show that RFID is essential for the competitiveness of Europe's economy as an RFID user. It is important to remember that the progress and effectiveness of RFID should not only be measured simply by the number of tags or amount spent on RFID, however it should be measured by judging how far the technology penetrates into daily life, and if new application fields are established through innovation.

Europe as an RFID Technology Provider

But Europe is not simply an RFID user; its companies are also important suppliers of the technology – both hardware and software, as well as being important system integrators. Unlike other high-tech areas such as the computer industry, Europe plays an important role in almost the entire supply chain, from integrated circuits to manufacturing systems, from smart labels to tag readers. Again, only a few figures which show the economic importance of the European RFID industry exist. The above-mentioned German study predicts a turnover of € 1.4 billion by 2010 (Bovenschulte et al. 2007) for German RFID companies. A European figure should reach at least three times this amount. Although it is clear from the figures that the economic benefits for suppliers are noticeably lower that those for the users, it is important that Europe maintains a strong position as an RFID technology provider. An RFID market analysis study concluded that consistent growth in the RFID market will lead to heightened competition as more global players try to enter the market; therefore it is important that smaller players try to claim as much market share as possible (Eschinger 2008). Strong European RFID users need strong European RFID companies who are aware of their specific requirements when shaping this technology.

All this is situated amongst fierce global competition. As with other ICT industries, the US have an important position with respect both to RFID applications and technologies. Many South-East Asian countries, currently Japan and South Korea, in the future the People's Republic of China; are pushing forward the use and the domestic production of RFID systems.

Societal Advantages and Challenges

Besides its positive economic effects, radio frequency identification may also bring along societal advantages. Examples are the detection of a broken cool-chain in the food sector, the improved identification of patients, medicines and surgical instruments in hospitals, which prevents mistreatments, and faster services at border control based on electronic identity cards and passports. First studies from Maghiros (2007) or from Bizer et al. (2006) show these benefits in detail.

However, as RFID leads to a more networked world, sometimes dubbed the Internet of Things, there are potential risks that stem from the envisaged ubiquity of radio frequency tags. This comprises for instance, the endangered privacy of personal data, the loss of security for information in open systems, and environmental pollution by hazardous tag components in mass applications. A comprehensive approach to a European RFID innovation system must also meet these challenges.

A Contribution to the European Debate on RFID

With the RFID Consultation Process the European Commission has started a debate on these issues. The first milestones of this debate were the outlines in the Communication by the Commissioner for Information Society and Media – Vivian Reding, in March 2007 (CEC 2007), and the "RFID Policy Paper" that was published by the Commission and the German Federal Ministry of Economics and Technology in May 2007 (BMWi 2007). The Communication and the Policy Paper state alike that RFID is a promising technology that should be supported by a coordinated policy in the areas of R&D funding, European and worldwide radio regulations, and the governance of future RFID networks.

This book aims to contribute to this debate from the specific perspective of European companies that use, develop and deploy RFID technologies in their daily business. Thus, the focus of this book is not put on visionary application scenarios, but on today's challenges for RFID technology providers and users as well as on pragmatic approaches to meet these challenges. As it often happens with new technologies, challenges to the adoption of RFID are not just technological issues, but political and societal issues as well regarding privacy, health and environment.

Therefore, this publication summarises the results elaborated within the European coordination action project "CE RFID". CE RFID was funded within the 6th Framework Programme of the European Commission (Information Society and Media Directorate-General) and stands for "Coordinating European Efforts for Promoting the European RFID Value Chain". It is a European industry consortium designed to assess the challenges to RFID development from an industry point of view. This publication provides information about the current status of RFID in Europe, and gives definite recommendations for decision makers on how an appropriate environment for RFID development in Europe could be created. Therefore this book is aimed at decision makers from politics, economics, science, and administration, and provides helpful guidance on different subjects in the field of RFID.

Radio frequency identification is now a well-known technology. As there are a number of excellent introductions to RFID, this book abstains from simply more descriptions of RFID technology and applications but aims at giving recommendation for supporting the technology. This is how it differs from the valuable books by Finkenzeller (2003) which involve specific technical details; by Fleisch and Mattern (2005), which puts its focus more upon applications; and by Bullinger and ten Hompel (2007), which covers current technologies as well as economic and logistical aspects.

1.1 The Technology Diffusion of RFID and Specific Challenges

The further development of RFID in Europe can be supported in different ways, and by depicting an adoption model we wish to highlight the current challenges for RFID technology providers and users, which prevent RFID from developing further. We will then discuss later, how these challenges could be overcome. By using the well established model of Rogers (1962) as a base, a diffusion model can be developed, which illustrates the different phases along the RFID adoption curve. Time is depicted along the x-axis, and the accumulated use of technology on the y-axis. The first phase is characterised by a flat increase of the curve, the

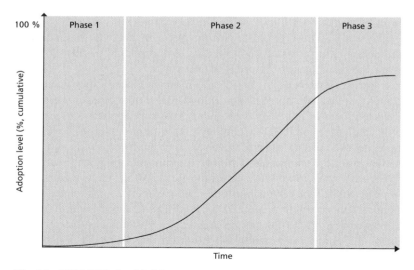

Fig. 1.1 RFID Diffusion Model

1.1 The Technology Diffusion of RFID and Specific Challenges

second phase shows a steep increase and the third phase again shows a flat increase. During the first phase, early adopters begin to use the technology. This group contains firms who act quickly, perhaps have a sufficient R&D budget, or perhaps are simply looking to create a competitive advantage by adopting technology which other competitors do not have. During the second phase, the technology turns into a mass technology and many users start locking in. This leads to a maturity phase at the end when laggards hop on and use the technology.

The adaptation of Rogers' model will not result in a predictive model for implementing RFID. However, the technology adoption model provides a method to highlight the need for action in order to overcome the barriers preventing the technology from advancing along the curve and integrating itself into society. Due to the differing characteristics of each adopter group, one can see that in order to further support technology diffusion, one must first meet the requirements of the respective adopter group. These characteristics have been formulated by listening to the feedback from the workshops carried out in CE RFID.

The shape of the S-curve not only depends on the user groups, but on many different topics concerning the technology itself, e.g. how innovative and advantageous it might be for users, or what alternative technologies exist. Furthermore, external factors such as institutional conditions also shape the form of the curve.

Studies have shown that RFID is currently reaching the end of the first phase of adoption, and continuing a strong level of growth away from pure research pilots and areas where businesses are trying to increase their competitiveness (Eschinger 2008b). The most critical step is the transition from phase 1 to phase 2. During this time the market decides whether an innovation will be successful or not. It is therefore important to analyse the barriers to phase 2's entry, which currently exist, to further support the development of RFID in Europe. From expert discussions and general research, the following critical success factors can be summarised.

Table 1.1 Adoption stages

	Phase 1	Phase 2	Phase 3
Users	*Early adopters* Innovative companies Companies with large R&D budgets Risk taking companies Technology developers	*Majority* Require proof that technology is reliable and produces positive ROI Require detailed information and guidance for use of technology	*Laggards* Smaller businesses with little or no R&D budget Risk averse companies Less technologically advanced firms
Technological Status	Low level of technological maturity R&D and field testing is still needed Development by companies who want 1st mover advantage Relatively high costs	Technology is more mature Increased user friendliness Widespread use of technology Sunk costs	Availability of Plug & Play solutions Low price of equipment

- **Standards**: In order to support technological developments, a reliable institutional framework is essential. Standards are one necessary requirement for creating fair, accessible and effective markets. This holds true for RFID as well, whereas the regulation of radio frequency bands used by RFID plays an equally important role alongside data and communication standards.
- **Guidelines**: Technological development and a reliable institutional framework which meets the needs of users are prerequisites for technology adoption. To facilitate and accelerate the adoption process, adequate information for potential users is necessary. RFID users are not usually experts in the fields of standards, radio regulation or law. They need best-practice examples, return-on-investment statements, and guidance for implementation. Potential users often criticise a lack of information, specifically guidelines, which are needed for implementation.
- **Legal framework**: Not only the information available for interested users and technically oriented standards and radio regulations can support the development of the RFID technology; also legislation is an important issue as it can either support or also impede technological growth. Therefore there is a need to analyse legal aspects of RFID. This comprises privacy, RFID in workplace as well as health and environmental protection.
- **Technology roadmap**: Although RFID use is spreading in Europe, it is surprisingly difficult to find a detailed overview of information regarding the properties and performance of RFID systems in Europe, and to produce realistic and profound estimations regarding future developments in hardware, software and applications fields. However, many technology users and suppliers need a reliable forecast of how RFID will develop and which will be the most important development tasks.
- **Research policy**: Supporting and accelerating technology adoption can be done in many different ways as described above. Besides institutional and informational support, assistance for research activities is needed as well. Although no longer an emerging technology, the broad scope of potential RFID applications and the number of open issues which are on their way to full implementation calls for substantial R&D effort in many domains. Regarding the fierce international competition the building of an effective European innovation system for RFID is of great importance.

We consider these five topics as critical success factors which must be taken into consideration if RFID is to develop its full potential in Europe. These critical success factors will be discussed in this book in their corresponding chapters, stating what needs to be achieved in the different areas for RFID technology to be a success.

1.2 Outline of the Book

Alongside the above mentioned five critical success factors, there are two additional main factors which influence the shape of the diffusion curve and which

1.2 Outline of the Book

will be discussed throughout the whole book. The first factor is the vast variety of RFID applications, which makes it difficult or even impossible to draw a diffusion curve for "RFID Technology" in general. The second factor is that there are many different stakeholders who will have an influence on its overall success and who interact with each other and use the technology or might be affected by others using the technology. In order to effectively structure the discussion on RFID from these two perspectives, appropriate models were created, each classifying RFID stakeholders and RFID application fields. These models are further explained in Chap. 2.

Chapter 3 introduces the main standardisation bodies and radio regulation authorities which act on international, European or national level. In addition, the most important RFID-related standards and radio regulations will be presented and analysed, distinguishing between application specific standards and application-independent standards and regulations for frequency bands, air interfaces, health and safety, privacy and environment. A further subsection deals with intellectual property rights (IPR) and the influence of different IPR models on standardisation. Chapter 3 closes with recommendations on RFID standards and radio regulations for European and global stakeholders.

Chapter 4 surveys and analyses the existing guidelines on RFID implementation. This work is based on a categorisation of interest groups within the organisation of RFID users and their specific demand for information. An important result of this analysis is a checklist, which can help companies and organisations to set up new guidelines – the RFID Implementation Checklist. The section closes with recommendations for authors of future RFID guidelines, in order to reduce the existing information deficit.

As it is true for any other technology, RFID must comply with national and European laws that implement for instance the protection of privacy, health and environment. These legal aspects are discussed in Chap. 5. Firstly, privacy aspects, (which are almost always addressed when discussing RFID legislation, even if personal data is not involved) will be examined in detail from a legal viewpoint with respect to the current data protection laws in Europe. In addition, we will discuss issues such as RFID in workplaces, the impact on health or on the environment and how these issues are covered by existing legislation. In addition, the existing IPR framework (Intellectual Property Rights) is analysed followed by a discussion on RFID governance. To conclude the legally orientated discussion, recommendations for further regulatory actions are drawn for each topic.

Chapter 6 begins by highlighting key technological bottlenecks in the field of RFID in Europe in various individual areas. It then goes on to use this analysis to outline a roadmap which defines mid and long-term RFID development targets in Europe, including estimations about developments in the field of hardware, software and applications. This roadmap serves as a guide which highlights the key targets which need to be achieved over the next fifteen years, in order to promote the advancement of RFID technology in Europe.

Chapter 7 will assess and analyse the present RFID Research and Development (R&D) policy in Europe by analysing the present implementation of European and national RFID R&D programs. This is carried out alongside a set of criteria, such as the accessibility of the programs or their coverage of innovation phases. The section closes with recommendations as to the content and the method of how to implement an optimised RFID R&D policy by asking how Europe could set the scene to optimise the return on subsidy money spent on both a national and a European level.

Chapter 2
Framework for the Classification of RFID Applications and Stakeholders

As discussed in the previous chapter, technology growth can be influenced by many different factors which have a lot of interdependencies. Besides the main topics that will be analysed in this book, there are two more themes which influence the diffusion of the technology and which will be discussed throughout the whole book. The first theme is the technology itself. A vast variety of RFID applications exist, which makes it difficult to predict the development of RFID Technology in general. The second theme presents a societal point of view; different people who interact with the technology will have an influence on its overall success. These people are named as stakeholders throughout this book. In order to incorporate these two topics, two models were created. One aims at structuring RFID applications (the RFID Reference Model), and the second aims at structuring stakeholders (the RFID Stakeholder Model).

2.1 The RFID Reference Model

RFID is a technology which permeates into many application fields. It was originally developed for use in World War 2 to distinguish between friend and foe aircraft. Today, RFID technology is used in various areas of our daily lives. When an application field is so broad, it is helpful to have a model which can help classify applications so that a clear understanding can be achieved. The RFID Reference Model is a first approach in this direction. The RFID Reference Model has been developed as an attempt to structure the different application fields in order to be able to evaluate requirements in terms of standards, technological components, and data protection for individual applications (Gampl et al. 2008b).

The first step when developing the RFID Reference Model was to create a list of RFID applications, and then to discuss them with various RFID experts. The idea was to create a list of applications that is as comprehensive as possible. These applications have then been merged into RFID application fields. The eight different application

10   2 Framework for the Classification of RFID Applications and Stakeholders

RFID-Application Field		Subcategories
Tagging with Reference or Potential Reference to Individuals	H. Public Services	HA. – Public Service Maintenance HB. – Road Tolling Systems HC. – Banknotes HD. – ID Cards and Passports
	G. Sports, Leisure and Household	GA. – Sports Applications GB. – Rental Systems GC. – Smart Games GD. – Smart Home
	F. eHealth Care	FA. – Assistance for Disabled FB. – Hospital Management FC. – Implants FD. – Medical Monitoring FE. – Smart Implants
	E. Loyalty, Membership and Payment	EA. – Loyalty Cards EB. – Membership Cards EC. – Contactless Banking Cards ED. – Payment and Advertising via mobile phones
	D. Access Control and Tracking & Tracing of Individuals	DA. – Ticketing DB. – Access Control System DC. – Animal Tracking DD. – Personal Tracking
Mainly Object Tagging	C. Product Safety, Quality and Information	CA. – Fast Moving Consumer Goods CB. – Electronic Goods CC. – Textile Goods CD. – Fresh/Perishable Foods CE. – Pharmaceutical CF. – Customer Information Systems
	B. Production, Monitoring and Maintenance	BA. – Archive Systems BB. – Asset Management (incl. Environmental Monitoring) BC. – Facility Management BD. – Vehicles BE. – Airplanes BF. – Automation/Process Control BG. – Food and Consumer Goods
	A. Logistical Tracking & Tracing	AA. – Inhouse Logistics AB. – Closed Loop Logistics AC. – Open Logistics AD. – Postal Applications AE. – Dangerous Goods Logistics AF. – Manufacturing Logistics

Fig. 2.1 RFID Reference Model

fields (A–H) are shown at the top of the diagram (Fig. 2.1). The first three application fields (A–C) address mainly object tagging applications, whereas the latter fields (D–H) address situations where the tags have an immediate reference or potential reference to individuals. The application fields are then subdivided into different subcategories. These are listed alongside their respective application field.

These application fields explain in which areas RFID is used. The first area, "Logistical Tracking & Tracing", contains applications where RFID is used solely for the location and identification of goods and returnable assets. This can be for example, an RFID system within one company, which acts as an in-house solution, or an RFID system that is used along a supply chain. "Production, Monitoring and Maintenance" defines areas where RFID is used alongside smart systems to support the production, monitoring and maintenance of goods and processes. This can be for example, in the automotive industry when producing cars along assembly lines almost automatically, or in food producing companies for monitoring the production process, or in the aviation industry for maintenance processes. "Product Safety, Quality and Information" includes areas where RFID is used to monitor quality and safety, for example with integrated sensors and supports anticounterfeiting of products. "Access Control and Tracking & Tracing of Individuals" covers areas where RFID is used for identification and authorisation of individuals, for example more and more hotels use RFID technology to give access to the rooms. "Loyalty, Membership and Payment" defines areas where RFID is used alongside smart systems for authorisation purposes in multi functional applications. These are applications where the identity of the user (including the respective rights to different actions) is saved on the tag. Areas where RFID is used for hospital administration and in the health care industry are defined under the "eHealth Care" section. This could be for example tagging beds, blood or patient reports for increased safety and security, and improved organisation. "Sport, Leisure and Household" defines areas such as car rental services, libraries, and other leisure areas. It also covers applications such as the "Smart Home" which is always discussed vividly when it comes to "smart" household appliances. "Public Services" define the areas where RFID is used in the public sector, areas such as road tolling systems or health insurance cards or identification documents.

By using these application fields, the RFID Reference Model acts as a useful reference tool which helps to structure the discussion on RFID. It has been published and updated four times to allow new applications to find their way in and to give room for improvements and new developments where necessary. It can be downloaded from the project website (www.rfid-in-action.eu).

Since the RFID Reference Model was published as a first version, it has received considerable attention from both researchers and companies with activities in the field of RFID. The Institute of Computer Science and Social Studies at the University of Freiburg has for example used the RFID Reference Model to conduct a study on the current status and future plans regarding the implementation of RFID in companies in Germany. According to this survey (283 companies answered), 102 companies already use RFID and 45 more plan to implement an RFID system until 2009. These 102 companies state that they have in total 492 different applications (Strüker et al. 2008). When breaking down these applications into the eight application fields according to the RFID Reference Model, the majority (326 of 492 applications) can be found in the first three fields ("Logistical Tracking and Tracing", "Production, Monitoring and Maintenance", and "Product Safety, Quality and Information").

2.2 The RFID Stakeholder Model

It was important to define the term RFID stakeholder for this project, as the term has different meanings depending on which group uses it. Stakeholders can be seen in both a business and a societal context as entities who may affect or who may be affected by respective actions. In the case of RFID technology, stakeholders are not only groups who provide or use the technology, but also groups who might be affected by others who implement and use the technology.

For decision makers and policy makers it is important to have a clear idea of RFID stakeholders. In order to achieve this, a model has been developed to clarify the term: The RFID Stakeholder Model (Fig. 2.2). The inner circle shows RFID stakeholder groups. These are both RFID users (RFID end user companies and RFID technology suppliers) and groups that might be affected by companies using RFID technology or that show interest in this topic. The research and development group consists of R&D organisations, both public and private, such as universities and research institutions, working in the fields of basic or applied research. Business associations consist of industry organisations, for example with a general technological focus or perhaps an RFID specific focus. Government and governmental organisations contain political groups from both national and European level. Standardisation organisations include international organisations such as ISO (International Standardisation Organisation) and industry driven organisations such as EPCglobal. Quasi-autonomous and non-governmental organisations are for example government appointed agencies, consumer organisations or trade unions.

The RFID Stakeholder Model structures the discussion about RFID stakeholders. Having a clear picture of who the relevant RFID stakeholders are is essential e.g. for a company or business association implementing an RFID system in order to give them adequate, useful, and understandable information. Depending on the applications, different stakeholders are affected and need to be informed by addressing their specific informational needs.

2.2 The RFID Stakeholder Model

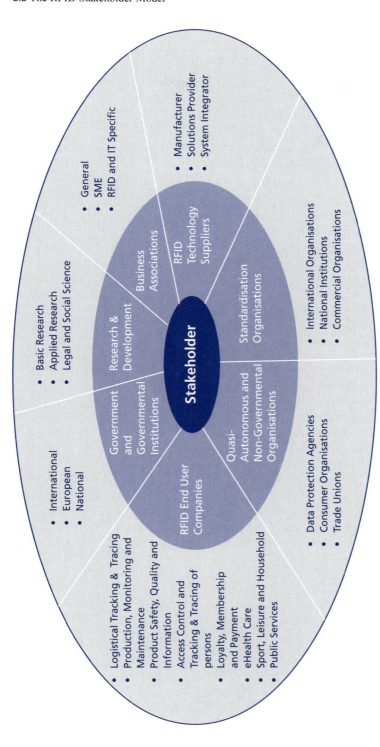

Fig. 2.2 RFID Stakeholder Model

Chapter 3
Standards

Standardisation is a process of developing and agreeing upon technical standards and making them a vital instrument for new technologies to avoid an uncontrolled growth of research results. Standards regarding RFID concern frequencies, communication, data, networks, safety and applications, and need to be internationally coordinated. One main challenge which organisations encounter with RFID implementations is the lack of industry-wide standards (Goel 2007). Therefore the following survey of RFID standardisation processes also includes recommendations for future activities.

This chapter deals with the topic of RFID standardisation. The main objective of this section is to collect and analyse all kinds of information regarding RFID standards. This includes

- RFID frequencies and radio regulations,
- RFID communication standards,
- RFID data standards,
- RFID network standards,
- RFID security and privacy standards and
- RFID application standards.

The analysis of these standards has taken place with respect to overlaps, inconsistencies, sector specific standards and gaps within standards. After the analysis of the different standards, recommendations for European RFID standardisation activities have been developed. Some of these recommendations have already been adopted, and some standardisation activities are currently ongoing. These recommendations will allow the creation of supportive standards for RFID, thereby allowing a controlled growth of the technology, and interoperability of systems throughout Europe. By doing this, collaborative efforts can be coordinated, thus allowing the development of the technology to advance further along the adoption curve and into European society.

To establish a specific technology it is important that there are several standards available on the market. However, over-standardisation as well as under-standardisation might prevent the establishment and the success of a technology. Both might prevent the establishment and the success of a technology. In breakthrough environments in particular, and with immature and rapidly developing technologies, over-standardisation can hinder innovation and under-standardisation can impede investment into the technology. Under-standardisation can also force regulation to control social or environmental effects resulting from improper standards. In addition to this, in certain application areas in which there is only a small number of players, proprietary standards might be created in order to cover the costs for their elaboration. This might result in the creation of entry barriers for other players and in misleading standards.

3.1 Standardisation Organisations and Processes

Standard development organisations (SDOs) play a major role in regard to design and relevance of international and national RFID standards and regulations. There are international as well as national organisations, each with different tasks. According to their responsibilities, their members and the type of standards they develop, SDOs can be divided into three major subgroups:

- Basic rule setting organisations
- Standard development organisations
- User and industry organisations.

3.1.1 Basic Rule Setting Organisations

Basic rule setting organisations like the ITU (International Telecommunications Union) or the European Parliament have decisive and innovative roles. Deliverables from these organisations usually serve as a basis for mandatory and harmonised regulations. They are defining a framework in which standards can evolve. Often they have a legally binding character. These organisations discuss and coordinate the basic regulations of global telecom networks and services. All other standards are orientated towards these rules. Basic standard setting organisations usually operate on an international level.

Deliverables made by the European Commission are usually recommendations and directives with a legislative character. They are given to the governments of the Member States, which should implement these recommendations into national law. According to this there are several standards which are influenced by these laws, e.g. a device must comply with several standards to receive the CE labelling.

The ITU-R (International Telecommunication Union – Radio Communication Sector) is a sub-group of the ITU. This group is responsible for the management of radio-frequency spectrum and satellite orbits.

3.1.2 Standard Development Organisations

Standard development organisations (SDO) develop standards for RFID (and most often for other technical sectors, too) and operate within the frameworks elaborated by the basic rule setting organisations. Sometimes they are mandated by the rule setting organisations, as it is the case for ETSI which is mandated by the European Commission for a number of RFID standards.

Standard Development organisations operate on different levels. ISO (International Standardisation Organisation) or ETSI (European Telecommunications Standardisation Institute), for example, operate internationally, while organisations like DIN (German Institute for Standardisation), AFNOR (French Association of Standardisation) or BSI (British Standardisation Institute) only develop standards for use on a national level, or on a specific topic. SDOs use the laws and regulations made by basic rule setting organisations as a starting point for their work. With their standards they define the specifications of a system and the environment around it.

ISO (International Organisation for Standardisation) is one of the world's largest developers of standards. The principal activity of ISO is the development of technical standards. The standards are used by various groups: industrial and business organisations of all types, governments and other regulatory bodies, trade officials, conformity assessment professionals, suppliers and customers of products and services in both the public and the private sector. The ISO standards have an important economic and social repercussion. They try to contribute to making the development, manufacturing and supply of products and services more efficient and safer, the trade between countries easier and fairer and they aim at providing governments with a technical basis for health, safety and environmental legislation. The most important ISO RFID standards are standards which describe the over-air-interface between a reader and a tag, and standards which describe the management of data.

The task of ETSI is to produce telecommunication standards for today and for the future. It is officially responsible for the development of standards for Information and Communication Technologies (ICT) within Europe. These technologies include telecommunications, broadcasting and related areas such as intelligent transportation and medical electronics. ETSI is developing a wide range of standards and other technical documentation as Europe's contribution to worldwide ICT standardisations. These standards are close to market needs and have a broad acceptance. The primary objective of the organisation is to support the global harmonisation by providing a forum in which all players can participate actively. ETSI reports to the European Commission and is officially recognised by the secretariat of the European Free Trade Association (EFTA). The most important RFID-related ETSI Standards are radio regulations. Two different types exist: radio regulations for short range devices in general, and radio regulations specifically for RFID devices.

3.1.3 User and Industry Organisations

User and industry organisations are usually subscriber-driven organisations. The developed standards are oriented towards the interests and requirements of the industry. The adoption of standards developed by industrial organisations is generally not mandatory. The most important organisation in this sector is the non-profit industry-driven organisation EPCglobal which was founded by the industry organisations GS1. GS1 itself is a merger of the three barcode bodies EAN International from Europe, Uniform Code Council (UCC) from the USA and Electronic Commerce Council of Canada (ECCC). EAN, UCC and ECCC started as organisations mainly driven by retail and the consumer goods industry. Today, GS1 aims to cover other sectors like pharmaceuticals or automotive as well. EPCglobal tries to further the development of industry-driven standards for the Electronic Product Code (EPC) as a successor of the conventional barcode. Standards developed by users and industries usually describe details of communication protocols and applications. EPCglobal has already developed a large number of RFID standards. The most important standards are the Gen2 Air Interface Protocol for operation in a frequency range from 860 MHz to 960 MHz, and the corresponding Gen2 standard for the HF interface. Alongside these two standards, many data and network standards exist which are becoming more and more important.

3.1.4 Business Models of Standardisation Organisations

Each of these different organisations has its own financing and business model. The most usual ways of funding are:

- Membership fees
- Grants
- Revenue from assets
- Fees for the use of identification numbers
- Selling of standards

ISO funding is based on subscriptions, which are paid by the national members. The subscription fees meet the operational costs of ISO's Central Secretariat. The subscription paid by each member is calculated in proportion to the country's gross national income and trade figures. However, the operations of ISO Central Secretariat represent only about one fifth of the cost of the system's operation. The main costs are carried by the member bodies who manage the specific standards development projects and the business organisations that provide experts to participate in the technical work. These organisations are, in effect, subsidising the technical work by paying the travel costs of the experts and allowing them time to work on their ISO assignments. Another source of revenue is the sale of standards. Other formal European standards development organisations like CEN (European

3.1 Standardisation Organisations and Processes

Committee for Standardisation) and CENELEC (European Committee for Electrotechnical Standardization) follow a way of funding similar to ISO.

Annual membership fees are the principal source of ETSI's income. The fees are usually related to the financial turnover of member companies or organisations. In the case of national administration membership (government ministries, etc.), the annual fee is related to each country's gross domestic product (GDP). Because the association is established under French law, ETSI is not permitted to make any profit, and any surplus of income over expenditure in a financial year is usually returned to members as credit to their membership fee for the ensuing year. Alongside membership fees, there are several other sources of income. As a European standardisation body, ETSI receives contributions from the European Commission and the EFTA Secretariat. The commission pays ETSI for mandated (contracted) standardisation work and for participation in special projects. Furthermore, ETSI receives income from commercial activities such as the selling of standards, and events and services to partner and outside organisations.

In contrast to other standard development organisations, EPCglobal/GS1 receives income through sales of identification numbers in the different organisation fields. Users of the GS1 standards have the opportunity to buy a specific range of numbers which they are allowed to use in their application. This is the same for RFID applications, where users can buy Electronic Product Code (EPC) numbers as well as for applications using the European Article Number (EAN).

One important factor for the adoption of a standard is the cost. The use of a standard should be as low-priced as possible. The costs for the use of a standard are not influenced decisively by the fees which have to be paid to the different organisations. Major costs are related to Intellectual Property Rights possibly held by a company for a certain standard. It will therefore be an important task to promote the creation of royalty-free standards and open standards. Everybody, no matter if the company is a member of the standard development organisation or not, shall have free access to developed standards.

It is important to bear in mind that it is possible for companies to participate in the work of European and national SDOs. One example of a formal standardisation organisation at a European level which allows the effective development of standards by offering participation for companies is ETSI as it is designed to allow an open level of participation for companies in the development of standards. In that way it can be ensured that there are no anticompetitive advantages for the organisations which are ruling the different SDOs.

> **Recommendation:** Support creating fewer but broader accepted standards

It is important to create fewer, but broadly accepted standards. This would provide more possibilities. If the complete set of regulations is given, technology providers as well as users of a product do not have any leeway to create their own solutions. If there are many standards for similar topics they need to decide on one and that would be an equal situation to having no standards at all. Unfortunately, there is still a lack of well-established standards. Many standards are in place; however, there is a weak agreement on which ones should be held up. This problem exists particularly with respect to data and network standards.

> **Recommendation:** Support basic standards like frequency allocations, air interface protocols and tag data specifications

Firstly there is the field of frequencies and air interface protocols. Only if frequency bands and air interface protocols are harmonised, will a complementing development of standards be achieved. Harmonised regulations for Europe will also strengthen the position of technology providers, who do not want to waste time, effort and money by developing different types of hardware, which are adapted to several local regulations. In this way globally harmonised standards and frequency regulations can help to reduce costs.

Secondly, there is the field of tag data. This especially concerns identifiers and codes. The compatibility between different identifiers and codes from different issuing agencies has to be ensured as it is implemented in the ISO Standard 15434, which describes the interoperability between different numbering systems such as GS1. Standards which describe the interoperability between different numbering systems have been compiled in the standards database by Walk et al. (2008).

> **Recommendation:** Support the development of royalty-free standards

A further point that should be discussed is the influence of IPR (Intellectual Property Rights) on standardisation. The influence of IPR on existing and future standards should be minimised. It should be ensured that the most relevant standards are free from any royalty payments. Only in this way is the market able to develop freely and without any hindrances. Currently, many standardised technologies and procedures are owned by individuals through intellectual property rights. Everybody who wishes to release a product which conforms to a standard must pay for it. In addition, guidelines on licensing conditions for standards related to IPR should be developed. More legislative questions regarding this topic are dealt with in Chap. 5.

3.2 Radio Spectrum Framework

RFID Systems emit electromagnetic waves; they can be viewed simply as radio installations. Through the operation of RFID systems, other devices and systems shall not be disturbed. Due to this fact, only frequency bands can be used for RFID

3.2 Radio Spectrum Framework

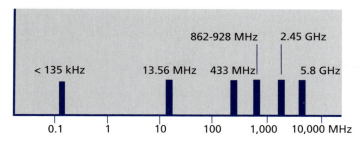

Fig. 3.1 RFID frequency bands

which are reserved for special industrial, scientific and medical applications. These are the worldwide available ISM frequency bands (industrial-scientific-medical).

Frequency Band 9–135 kHz

This LF frequency band is frequently used by different radio installations and is not declared as an ISM band. The propagation characteristics of these long wave signals allow an area with a radius of approximately 1,000 km to be covered with only a small technical effort. Fixed services, military, air and sea navigation systems or time signal systems usually use this frequency band.

The ETSI standard EN 300 330 describes the technical details and test methods of a device.

The ISO/IEC 18000-2 defines the air interface for RFID devices used in item management applications. It also describes some technical attributes and the communications protocol used in the air interface.

ISO 14223-1 specifies the air interface between the transceiver and an advanced transponder used in the radiofrequency identification of animals under the condition of full upward compatibility according to ISO 11785.

This frequency band is accepted worldwide. The fact that this frequency band is used frequently and that RFID systems can interfere with other devices may cause problems. The frequency 77.5 kHz is particularly critical, due to the time signalling system. The future German regulation for inductive radio systems (220ZV 122) will declare a protection area between 70 kHz and 119 kHz to avoid interferences with frequently used devices and applications. Lately the number of applications operating at this frequency has decreased, for example the German Weather Service (DWD) stopped sending weather information at this frequency band in 1996. The current frequency range is sufficient for RFID applications in this band. The band should be protected by frequency regulation authorities and therefore be available in the future.

Frequency Band 13.56 MHz

The conditions in this short wave band (13.553 MHz to 13.567 MHz) allow it to transmit information trans-continentally. The resource is used by different kinds of users, like press agencies and telecommunication companies. Typical applications for this frequency band are telecontrol systems, personal signalling devices, remote controls and demonstration facilities.

The ETSI standard EN 300330 defines technical details and test methods for devices operating in the HF frequency band.

The HF standard ISO 18000-3 provides physical layer, collision management system and protocol values for RFID systems in accordance with the requirements of ISO 18000-1. Furthermore, two modes of operation are described, intended to address different applications.

ISO/IEC 15693 specifies the nature and characteristics of the fields to be provided for power and bi-directional communications between vicinity coupling devices (VCDs) and vicinity cards (VICCs). Furthermore, it describes protocols and commands as well as methods for anti-collision.

The EPCglobal HF Generation 2 air interface protocol defines the physical and logical requirements for a passive load-modulated, interrogator-talks-first (ITF), radio frequency identification (RFID) system operating at 13.56 MHz frequency range. The system comprises interrogators, also known as readers, and tags, also known as labels. This standard will also be implemented in ISO 18000-3 as Mode 3. The standard employs the ISO/IEC 18000-6C logical layer to enable the code re-use for the logical layer. With certain novel antenna designs it is conceivable that ISO/IEC 18000-6C and 18000-3m3 might be able to be simultaneously developed in the same chip.

Current RFID standards are well qualified for applications with a limited number of transponders. For the future there is the need for a worldwide accepted HF high speed RFID standard, in order to realise high data rates.

Frequency Band 433 MHz

The frequency range from 430 MHz to 440 MHz is used worldwide by radio amateurs. They transmit speech as well as data and use them for communication with self-made satellites. Buildings and other barriers have a large influence in the form of strong reflection and attenuation on electromagnetic waves. Depending on the output, power and operating method distances between 30 km and 300 km can be achieved.

The ETSI standard EN 300220 describes the electromagnetic compatibility and radio spectrum matters (ERM) for Short Range Devices (SRD) used in the 25 MHz to 1.000 MHz frequency range with power levels ranging up to 500 MW.

ISO/IEC 18000-7 defines the air interface for RFID devices operating as an active RF Tag in the 433 MHz band used in item management applications. The standard also defines technical attributes and the communications protocol used in the air interface. ISO/IEC 18047-7 explains test methods for active RFID air interface communications at 433 MHz.

This UHF frequency band is only used by RFIDs for a few applications. The most typical application is the identification of containers. For the future there is no need to work on standards and changes for this frequency band.

Frequency Band 840–960 MHz

This frequency band is for RFID UHF applications. The frequency used is dependent on the availability in the different countries:

- China: 840–845 MHz and 920–925 MHz
- Europe: 865–868 MHz
- United States: 902–928 MHz
- Japan: 952–954 MHz

The European UHF frequency band has been approved for short range applications since 1997. Technical characteristics and test methods for short range devices operating in the UHF frequency band with a maximum output power of 500 MW ERP (effective radiated power) are specified in the standard EN 300220. This standard was developed by ETSI.

The ETSI Standard EN 302208 concerns the electromagnetic compatibility and radio spectrum matters (ERM) of RFID equipment operating in the band 865 MHz to 868 MHz with power levels up to 2 W ERP. This standard provides 15 UHF RFID channels with a bandwidth of 200 kHz in the range between 865 MHz and 868 MHz. From these 15 channels only 10 channels are high power channels and can operate with a maximum output power of 2 Watts ERP. The tag backscatter is located in the same channel as the reader-to-tag signalling. This procedure requires a listen-before-talk-system. A problem of this system is, however, that only a maximum number of ten readers can operate simultaneously within an environment. Another problem is that listen-before-talk requires a special hardware which makes the readers more expensive.

The technical specification TS 102 562 was published in March 2007. This document provides some regulations for the implementation of the dense reader mode in conformance with the current European UHF standard. It describes a new four channel plan and methods for the synchronisation of the listen-before-talk-process.

The European Commission has confirmed its objective to harmonise the European UHF frequency spectrum. In November 2006 the Decision 2006/804/EC was adopted. This decision makes it mandatory for all European Member States to make a unique frequency spectrum available within 6 months. The spectrum parameters are based on the ERC/REC 70-03 and on the ETSI standard EN 302 208.

A new version of the ISO 18000-6 standard is currently under development. It describes parameters for the air interface communications at 860–960 MHz. The EPC Class1 Gen2 standard is commonly known as the "Gen2" standard. This standard defines the physical and logical requirements for a passive-backscatter, interrogator-talks-first (ITF) RFID system operating in the 860–960 MHz frequency range. The standard is corresponding to the ISO 18000-6c standard.

The UHF frequency band is an important frequency range for logistical applications. Unfortunately there are currently no unique regulations across Europe. A first step was made through the EC Decision 2006/804/EC. This decision makes it mandatory for all Member States to make available the frequency bands as they are described in the ETSI standard EN 302208. For the future there should be a larger UHF frequency spectrum. In the United States the spectrum has a bandwidth of 26 MHz while in Europe only a 3 MHz spectrum is currently available. A pure bandwidth can be compensated partly through a better utilisation of the available bandwidth. Further higher power levels could be very useful to improve the performance and reading rates of an RFID system. The maximum allowed power level of 2 Watts ERP causes only a limited transmission through the different materials.

Frequency Band 2.45 GHz

The frequency range between 2,400 GHz and 2,485 GHz is partly overlapping with frequency ranges of radio amateurs' applications and radar systems. The propagation characteristic for this UHF frequency band is nearly optical, comparable with the one of visible light. In other words, where light is barred, 2,400 GHz is too. Buildings and other barriers act as good reflectors and attenuate an electromagnetic wave very strongly during transmission. The most used applications next to backscatter systems are telemetric transmitters and wireless LAN.

ETSI EN 300440 describes the electromagnetic compatibility and radio spectrum matters (ERM) for short range devices operating in a frequency range from 1 GHz to 40 GHz.

The ETSI EN 300683 standard contains information about electromagnetic compatibility (EMC) of short range devices (SRD) operating on frequencies between 9 kHz and 25 GHz. This standard together with ETSI EN 300761 is intended to become a harmonised standard. EN 300761 describes the automatic vehicle identification for railways operating in the 2.45 GHz frequency range.

ISO/IEC 18000-4 was published in 2004. A new version of this standard is currently under development. It describes the automatic identification and data capture techniques for RFID systems for air interface communications at 2.45 GHz.

This frequency range is a very wide and frequently used ISM band. There are certain problems such as the frequent use of wireless LAN and Bluetooth applications. If an application intends to use the whole frequency band it is only possible to run active systems because of the low transmit power of 10MW. In the future there is a need to define special frequencies for RFID applications and WLAN applications and to separate them strictly from each other. Furthermore, the implementation of passive systems should be supported in this frequency band

Frequency Band 5.8 GHz

The ISM range from 5,725–5,875 GHz also partially overlaps other frequency bands of radio amateur applications and radar systems. Typical ISM applications are motion detectors, such as for door openers in warehouses, and backscatter systems.

The electromagnetic compatibility and radio spectrum matters (ERM) for short range devices operating in a frequency range from 1 GHz to 40 GHz are described in ETSI EN 300 440.

3.2 Radio Spectrum Framework

The ETSI EN 300 683 standard deals with the topic of electromagnetic compatibility (EMC) operating on frequencies between 9 kHz and 25 GHz.

The Standard EN 300 674 describes road transport and vehicle telematics operating in the 5.8 GHz ISM band. Part one of this standard describes general characteristics and test methods for road side units and onboard units, while part 2 specifies the requirements for the onboard units.

This frequency is used for specific applications. Currently, there is no need for action in regard to standardisation.

Recommendation: Ensure an appropriate radio spectrum framework

Presently there are some problems concerning radio spectrum in the European Union, such as the lack of harmonisation and full availability of "first generation" radio spectrum ranges, besides from the fact that they are considerably smaller than the available frequencies in other countries, which impedes the technological development of the EU. In addition, there is a growing need for a pan-European program to tackle radio spectrum matters.

Although some important legal instruments have recently been adopted in order to address those issues, they only partly meet the requirements for organising the sharing of radio frequency: the proposal for a directive repealing Council Directive 87/372/EEC, intended to "free" frequency bands that are currently reserved for GSMs, might not suffice for shaping a consistent plan for the upcoming radio spectrum necessities.

Several proposals were made for a better coordination of spectrum management at the Community level – a position iterated by the Commission as part of its telecom reform package suggested in November 2007, calling for a "common coordinated approach" to manage the digital dividend. Member States are urged to "set aside airwaves for EU-wide services and share spectrum on an EU-wide basis by clustering similar types of services into common spectrum zones. Clustering would also help to manage better interference between the main networks that support the services" (COM(2007) 700 Final).

Recommendation: Provide additional UHF spectrum

The RFID industry expects that the use of RFID in Europe will grow rapidly within the next 15 years. The commercial benefits of RFID are becoming more and more recognised and so the technology will be adopted by many new industries. A greater reading range, faster data rates and the use of sensors (e.g. temperature, pressure, etc.) within tags will be requirements of future RFID systems. An additional UHF spectrum should be provided to improve the performance of RFID systems and to improve Europe's situation on the RFID market.

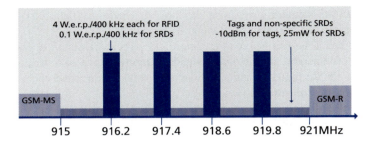

Fig. 3.2 Proposal for high performance RFID and SRD applications

In the long run an adequate UHF spectrum should be provided for high-performance readers and for low-cost readers as well.

There are already some activities to provide an additional UHF spectrum. ETSI is currently working on the technical report TR 102 649-2, additional spectrum requirements for UHF RFID, non-specific SRDs and specific SRDs. It is suggested that the frequency bands at 870 MHz to 876 MHz and at 915 MHz to 921 MHz should be made available for RFID systems and short range devices.

> **Recommendation:** Ensure the elimination of the necessity for listen-before-talk

ETSI is currently working on a revision of the Standard EN 302208 in which the listen-before-talk-process will be eliminated. The elimination of listen-before-talk is supposed to take place by the year 2008 – for instance, the United Kingdom has already eliminated it. After its elimination, changes in the ERC/REC 70-03 will be performed. As soon as this standard is regarded as published it has to be made sure that the regulations in the various Member States are transformed into national legislation.

3.3 Interoperability of Standards

There are a vast number of RFID-related standards and standardisation organisations. More than 250 standards describing RFID-related solutions have been established by around 30 different organisations. This large quantity of documents shows that RFID is of eminent advantage for a wide range of applications in different industries, as can be seen in the RFID Reference Model. It is essential that standardisation organisations cooperate, in order to avoid different solutions for equal requirements. As products can be used for many applications, the economy of scale can be leveraged.

3.3 Interoperability of Standards

The interoperability of standards plays an important role in the coexistence of technical solutions requiring the same resources like frequencies. Within the RFID product range the uniqueness of product codes has to be guaranteed by well defined rules. The development of RFID technology requires interoperable standards. In particular, given the globalisation of the economy, worldwide interoperability is necessary. It will, for instance, enable the use of RFID technology from the beginning to the end of the supply chain, regardless of the countries or parties (companies, authorities) concerned.

Although not primarily a legal issue, the European Commission should retain an active role in promoting the interoperability of RFID standards across national and regional boundaries as well as different industry sectors. In this respect, the GRIFS project and the Trans-Atlantic Economic Council initiative (including RFID as one of its "lighthouse" projects) lead in the right direction and should receive the necessary attention.

3.3.1 Product Codes

One of the big advantages of RFID over barcode technology is that many objects can be identified automatically at once without having the need for a line of sight. This means every RFID transponder communicating with the right air interface protocol coming into the reading range of the antenna is read by the reader. The relevant data standards defined by ISO and EPCglobal provide smart filtering methodologies which ensure that transponders which do not belong to the specific application are filtered out from data processing. A precondition for this functionality is the uniqueness of code for a single item.

> **Recommendation:** Ensure cooperation and data interoperability between different code issuing agencies

For EPCs, GS1 acts as the issuing agency for codes. For many end users, many synergies could be leveraged by building on the family of GS1 Identification Numbers, known collectively as the GTIN (Global Trade Item Number), GLN (Global Location Number), SSCC (Serial Shipping Container Code), GRAI (Global Returnable Asset Identifier), etc. GS1 assigns EPCs to end users and ensures that the numbering schemes guarantee uniqueness. Industries which do not make use of EPC have to be forced to use the relevant ISO standards to assign identifiers.

Even when RFID transponders are used in closed loop applications, it has to be considered that transponders which remain on an item may cause number collision problems in a downstream process outside of the closed loop. This could be avoided if operators are convinced to use standardised products which can be cost-saving and of higher quality at least in terms of costs when changing the supplier.

The potential risk of number collision will increase tremendously with the evolution of RFID applications. Therefore it is of importance to ensure RFID users make use of the relevant data standards, and it is mandatory to encourage industries planning to issue codes to adhere to the rules defined by the standards.

3.3.2 The Internet of Things

Over a decade ago, Mark Weiser, a US-American scientist in information technology, developed a groundbreaking vision of future technological ubiquity: increasing "availability" of processing power accompanies decreasing "visibility" (Weiser 1991). In his opinion, the most profound technologies are those that weave themselves into everyday life until they are indistinguishable from it. Early forms of ubiquitous information and communication networks are evident in the widespread use of mobile phones, even more so than of the Internet.

Today, developments are rapidly underway to take this phenomenon an important step further, by embedding short-range mobile transceivers into a wide array of additional gadgets and everyday items, enabling new forms of communication between people and things, and between things themselves. A new dimension has been added to the world of information and communication technologies (ICTs): from anytime, anyplace connectivity for anyone, we will now have connectivity for anything. Connections will multiply and create an entirely new dynamic network of networks – an Internet of Things, which is neither science fiction nor an industrial hype. It is based on solid technological advances and visions of network ubiquity that are being realised.

> **Recommendation:** Harmonise the standardisation for Internet of Things and RFID

One of these varied technological applications, which make integration into everyday life possible and even encourage it, is RFID. In order to connect everyday objects and devices to large databases and networks, a reasonably simple system of item identification is important. Only then data can be collected, processed and items may be tracked in real-time. RFID offers this functionality and is therefore seen as one of the pivotal enablers of the Internet of Things. In other words, from the total amount of technological innovations accounting for the Internet of Things, RFID is only a part. It is crucial for current development that RFID solutions are not bound to a prefabricated definition of the Internet of Things. Practical RFID solutions on many operational levels – national as well as international ones – are absolutely feasible on the basis of current standards and technical surroundings.

New standards will be developed in the upcoming years in order to pave the way for the Internet of Things. It will be eminent to consider existing applications and to offer possibilities for feature upgrades wherever possible. New standards shall not conflict with the operation or degrade the performance of existing installations. One example of proper handling is the definition of tag classes by EPCglobal.

3.3.3 Data Exchange and the Object Name Service (ONS)

Early applications of RFID predominantly include closed loop applications such as public libraries, access control, inventory management and monitoring and steering of production in the car industry to name a few important examples. Those applications do not have the need to transfer product related data between different companies and are implemented considering a well defined subset of existing RFID standards successfully. Supply chain management applications or e-government applications such as RFID in drivers' licences, passports or cash have one fact in common: they are communication solutions based on the exchange of data between different business entities.

One important challenge in convincing users to adopt emerging technologies is the question of data security and the protection of privacy. Concerns over privacy and data protection are often discussed, particularly as sensors and smart tags can track users' movements, habits and ongoing preferences. Invisible and constant data exchange between things and people, and between things and other things, will take place, unknown to the owners and originators of such data.

If RFID is understood as a pure "communication solution" where items communicate and data is exchanged inside of one entity, there are a number of applications, such as banks or hospitals, which show the practical relevance of the technology and that it is not necessarily bound to further development of the Internet of Things during the current development process.

EPCglobal has established the ONS (Object Name Service) as standard to provide an automated networking service to make product related information available on the internet. With the help of an ONS, one or more URLs are determined for an existing electronic product code (EPC) when reading an RFID tag with an RFID reader. These URLs contain further information. This may be an internet address of a producer of a certain product. When RFID is used along the supply chain, there can be several URLs, e.g. of the producer, the distribution centre or the shop outlets. ONS points the middleware to all EPCIS (EPC Information Services) servers, where the product's files are stored. The middleware retrieves the files, and the files' information on the products can be forwarded to a company's inventory or supply chain application. The EPCIS servers might be located with different participants of the supply chain. The emergence of ONS will very much depend on market needs and user acceptance.

Standardisation is an important precondition for the mass deployment and diffusion of any technology and this also applies to RFID. Nearly all commercially successful technologies have undergone some process of standardisation to achieve mass market penetration. Successful standardisation in RFID was initially achieved through the Auto-ID Center and by EPCglobal. Standards had been defined on how an ONS on the basis of purely technical-organisational procedures works from the EPC data with the URLs. With it, the competence for data security lies in the hands of the respective domain holder. It is obvious that by using this methodology there is a potential risk that highly sensitive information on products, involved companies or even individuals can be found. Another threat is that root data or product data is being faked for criminal reasons. Therefore proper security measures have to be arranged, in addition to ONS, in order to ensure privacy. One of the most popular options would be to set up VPN (Virtual Private Network) connections between companies which are involved in data exchange. The advantage is that the methodology is reliable. Currently activities are ongoing which are identified as discovery services with the goal to standardise superior methods that guarantee secure data transfers. Unfortunately, this requires high coordinative and administrative efforts. A further risk is the loss of system functionality for technical reasons.

> **Recommendation:** European IT companies must support the standardisation process

It is not acceptable that highly sensitive processes like the supply chain depend on the availability of Internet connections. Redundancies need to be arranged within the system. The ONS concept requires the availability of centralised root information. A hierarchical approach is used to assign ranges of EPCs to issuing agencies in a way that the uniqueness of a specific product's EPC can be guaranteed. Each issuing agency can assign EPC ranges to EPC managers. In turn EPC managers may assign EPCs to companies. Large enterprises may assume the role of an EPC manager. This way, issuing agencies and EPC managers are able to assign EPCs totally independently from the central system. Each assignee can decide whether the related EPC range should be stored in the root ONS or not. The minimum set of information is the segmentation for the assignment agencies.

Until now, EPCglobal has awarded the US company VeriSign a contract to maintain the root ONS only. This shall guarantee that no redundancies are available and the security measures for data transfer merely apply to US regulations. There is an ongoing discussion about the governance of the ONS and a second ONS platform has been set up in France. The European root ONS has been available since March 2008, a first step to improve the situation (see Chap. 5.5.2 for a more detailed discussion on the governance on the ONS and the future Internet of Things).

There is a lot of work to be done in order to define comprehensive systems and standards around ONS. EPCglobal has started to investigate the white spots in the architectural framework, which is for example the threat that root data or product data is being faked for criminal reasons. Proper methodologies need to be developed quickly and this is a good opportunity to enable European experts to participate in the standardisation activities around discovery services.

> **Recommendation:** Support the development of proper data exchange standards

It is a great challenge to establish sustainable application-independent data and network standards. As the characteristics of RFID components and the complexity of the system may vary significantly from application to application, the standards have to strike a balance between regulation and flexibility. The definition of the system architecture clustered in building blocks that are separated by well-defined interfaces is crucial. EPCglobal's system architecture takes into account that not all RFID projects are carried out on an international level and therefore need an extensive set of regulations for standardisation. EPC global also permits national solutions which are easy to scale because of existing international standards and which allow exceptions from the rule: the more complex the RFID solutions, the lower the standardisation's stage of maturity. It can be expected that further development of this concept will be linked more to the emergence of the Internet of Things than it currently is. RFID applications of today and the near future can be built without the usage of ONS. The recommendation therefore is to foster R&D activities to look for alternative concepts to fulfil market needs and to guarantee privacy as well as to synchronise standardisation activities accordingly.

3.4 Analysis of the Need for Application Specific Standards

In the area of RFID application standards, the RFID applications fields "logistical tracking & tracing", "production, monitoring and maintenance", "product safety, quality and information" and the selected application subcategories "access control systems", "personal tracking" and "rental systems" of the RFID Reference Model have been collected and analysed. All of the different RFID applications fields were described in detail in Chap. 2.1.

In the subsequent sections, we describe which applications specific standards have been found (published or under development) and conclude if those applications that do not have their own application specific standards would need application specific standards. The structure of this section will be similar to the RFID applications fields of the RFID Reference Model. For a more detailed overview, please refer to the analysis in the CE RFID Standardisation report (Walk et al. 2008).

3.4.1 Logistical Tracking and Tracing of Goods

General Observations

Some generic RFID application standards exist for supply chain applications developed by VDI (Germany) and EPCglobal (International). VDI standards contain requirements concerning transponder systems, i.e. requirements for usage, cost assessments, security aspects and the management of RFID projects. The EPCglobal standards contain a recommendation for the design and use of standard logistic labels, including data backup functions as well as test methodologies for tag performance of door portals and conveyer portal applications.

Table 3.1 Logistical tracking and tracing of goods overview

Special application standards related to one subcategory I – International Standards E – European Standards N – National Standards	In-house logistics	Closed loop logistics	Open logistics	Postal applications	Dangerous goods logistics	Manufacturing logistics
Freight containers: supply chain management		I				
Freight containers: coding, identification and marking		I				
Freight containers: identification and electronic data transfer		I				
Freight containers: environmental characteristics for electronic seals		I				
Returnable containers: application programming interface (API)		N				
Returnable transport items/units: requirements for RFID tags		I				
Returnable transport items/units: requirements for RFID systems		N				
Transportable gas cylinders: identification and marking		I				
Transportable gas cylinders: identification and marking		E				
Disposal logistics: methods of identification of waste containers and/or determination of the quantity of waste		E				
Disposal logistics: requirements for RFID systems for waste containers		N				
Disposal logistics: identification of waste containers by using of LF RFID systems		N				
Air transportation: identification and marking of container and pallets for air freight			I			
Air transportation: baggage tag design				I		
Air transportation: identification and marking of baggage				I		
Military consignment and asset: marking and tracking					I	

3.4 Analysis of the Need for Application Specific Standards

Nearly all applications in this area are related to an international business which has a strong need for international standards ruling basic conditions of logistical tracking and tracing as well as the data and interfaces for data exchange.

In-house Logistics

No application standards were identified in this area, because a solely in-house logistic system is set up purely for a company's internal business processes with no interactions outside of the company.

Closed Loop Logistics

Within this subcategory many application standards exist. The identified applications could be clustered into two groups of applications. Firstly there are many international and national standards for the use of containers, freight containers and transport items/units. Secondly there is a sub-group of standards for special applications. The standards are related to ULDs (Unit Load Devices) for air transport (international), transportable gas cylinders (international and European) and for waste containers for disposal logistics (European and national).

In the subcategory of "closed loop logistics" for example, the supplier logistics for manufacturing, it would be extremely useful if more international standards were created. In this area, many national standards as well as some international standards were identified. Thus the harmonisation of existing standards will likely be a very useful approach in order to establish more unified international standards such as those for returnable transport items, which are presently only available as national standards. Only in special cases such as for disposal logistics will it be sufficient when European or national standards are available. However, international standards will not be an obstacle as each European country has to deal with the same duties.

Open Logistics, Postal Applications, Dangerous Goods Logistics, Manufacturing Logistics

In these subcategories, only three standards were found. Two international IATA standards were identified, which concern baggage handling in the business of air transportation. One military standard was identified, which is used for the tracking and tracing of military consignments, which are classed under the "dangerous goods logistics" subcategory. No standards could be found in the subcategories of postal applications and manufacturing logistics.

In the subcategories "open logistics", "postal applications", "dangerous goods logistics" and "manufacturing logistics", which all belong to international business areas, it was clearly asserted that international standards would be very useful to assist these businesses.

3.4.2 Production, Monitoring and Maintenance of Goods and Processes

General Observations

No generic RFID application standards for production, monitoring and maintenance applications could be identified. Due to the fact that the areas of production, monitoring and maintenance are very wide spread and complex businesses, it will be neither possible nor necessary to have standards to cover the complete application area.

Archive Systems, Asset Management, and Facility Management

No application standards were identified in these subcategories. The subcategories "archive systems", "asset management" and "facility management" possess specialist requirements, and their RFID solutions are isolated which means that common standards are only needed to govern general issues such as data and interfaces for data exchange.

Table 3.2 Production, monitoring and maintenance of goods and processes overview

Special application standards related to one subcategory I – International Standards E – European Standards N – National Standards	Archive systems	Asset management	Facility management	Vehicles	Aeroplanes	Automation/process control	Food and consumer goods
Automotive: common requirements on RFID				I			
Automotive: parts identification and tracking				I			
Automotive: methodology for the use for tire and wheel label				I			
Aerospace: passive RFID for parts and systems					I		
Aerospace: requirements on passive UHF RFID for aircrafts					I		
Aerospace: requirements for part/product identification and traceability schema for life cycle management					I		
Aerospace: passive UHF RFID for identification of parts and their traceability					I		
Semiconductor industry: RFID air interface for production and material handling equipment						I	
Production: data carriers for tools and chucking devices						N	
Maintenance: RFID-based wireless smart transducer interface for sensors and actuators						I	

3.4 Analysis of the Need for Application Specific Standards

Vehicles

Three different international AIAG (Automotive Industry Action Group) application standards could be identified within this subcategory. The first standard specifies common requirements of the automotive community regarding the use of RFID technology. The second standard specifies the application concerning the use of RFID for spare-part identification and tracking. The third standard specifies the use of RFID for product identification of tires and wheels. Within the subcategories "vehicles", the international standardisation process is already ongoing and is co-ordinated by industrial organisations.

Aeroplanes

A group of international application standards (Society of Automotive Engineers – SAE, International Organisation for Standardisation – ISO and Air Transport Association – ATA) were identified within this subcategory. The existing standards (SAE and ISO) specify the use of RFID for identification and tracking of aircraft as well as the technical requirements on RFID technology for this application. Additionally, the ATA standards (presently under development) specify the use of RFID for enhanced maintenance aspects like the tracing of aircraft parts during their lifecycles. As with the subcategory "vehicles", the international standardisation process is already ongoing to create standards for the "aeroplanes" subcategory.

Automation/Process Control

National (German DIN) and international application standards (IEEE and SEMI) could be identified within this subcategory. The identified standards have application areas equal to those for manufacturing/production and maintenance. Furthermore, a trend for the use of RFID transponders of type or rating plates on industrial equipment was identified. This trend aims to advance maintenance processes so that the development of according standareds will help to ease the application. In the subcategory "automation and process control" common standards are not likely to be achieved because the area is too heterogeneous, i.e. most of the applications are isolated and therefore need special solutions.

Food and Consumer Goods

No application standards could be identified in this subcategory. Due to the fact that the application area of "food and consumer goods" is usually an international business such as that for multimedia goods, it would be helpful if international standards were available.

3.4.3 Product Safety, Quality and Information of Goods and Processes

General Observations

Some RFID application standards for the field "product safety, quality and information" were identified. Some generic international standards for item management and some national standards for anti-theft systems and for electronic article surveillance are available. These national standards of the German company VDI are internationally used, as there are no comparable and suitable international standards available. In the area for customer information systems there are no generic standards available.

Because this application area is generally an international issue, it would be useful to have international standards for the complete application area to rule basic aspects of this business like data and interfaces for data exchange.

In regard to anti-theft systems and electronic article surveillance which are relevant to all subcategories, there are presently only some national standards available. In this case it would be very useful to have international standards which could be based on already established national standards.

Table 3.3 Product safety, quality and information of goods and processes overview

Special application standards related to one subcategory I – International Standards E – European Standards N – National Standards	Fast moving consumer goods	Electronic goods	Textile goods	Fresh/perishable foods	Pharmaceuticals	Customer information systems
Fast moving consumer goods: selection/integration of safety mechanism labels	N					
Fast moving consumer goods: selection/integration of safety mechanism labels		N				
Textile goods: requirements for the use of HF and UHF RFID systems for textile applications			N			
Textile goods: application instructions for clothes, shoes, leather goods, home textiles			N			
Fresh/perishable foods: traceability within agricultural and food branches				N		
Fresh/perishable foods: tracking and tracing of water bottles				N		
Fresh/perishable foods: requirements for cool chain applications				N		
Fresh/perishable foods: requirements for beverage logistics				N		
Fresh/perishable foods: perishable cargo handling in aviation				I		

3.4 Analysis of the Need for Application Specific Standards

In all subcategories only national standards are defined when there are any standards available at all. The only exception is the subcategory "fresh and perishable foods", for which an international IATA standard is available because of aviation issues.

In addition, for all of the subcategories, a strong need for international standards was identified. In all subcategories the reduction and prevention of plagiarism of products is a very urgent and important issue.

Fast Moving Consumer Goods

In this subcategory only one national standard of the German VDI was found which supports the selection and integration of safety mechanism labels.

Electronic Goods

In this subcategory only one national standard of the German company VDI was found which supports the selection and integration of safety mechanism labels.

Textile Goods

In this subcategory only some national standards of the German company VDI were identified, which describe the requirements as well as the application of the RFID technology in the area of textile goods.

Fresh/perishable Foods

Four national standards from France, USA and Germany have been identified which deal with specialised issues like agricultural goods tracing, water bottles or beverage logistics. One of the national standards of the German company VDI and the international guideline of IATA which deals with cool chain handling in general and especially in air transportation was identified.

Pharmaceutical

No application standards were identified in this area.

Customer Information Systems

No application standards were identified in this area. Within the emerging application area of the subcategory "customer information systems" no standards are available at present. However, a very strong need has been identified for international standards to support the establishment of these applications.

3.4.4 Access Control Systems, Personal Tracking, Rental Systems

General Observations

No general application standards were identified in this application field.

Access Control Systems and Personal Tracking

No application standards were identified for these subcategories. The RFID requirements within the subcategory "access control systems" are so broad that it does not seem to be possible to cover all of its aspects with generic standards. This is caused by the fact that many of these applications can only be found in local regions, and are restricted to companies and administrations. However, it may be helpful if international generic standards would be available to harmonise the technology in order to ease their application. In the subcategory "personal tracking" there is no need for international application standards because applications like tracking of children in leisure parks or in school areas, to find them if they are missing, are local and closed applications.

Rental Systems

In the subcategory of rental systems, only one international standard of ISO was identified. The standard defines the data model for the use of RFID technology for libraries. The subcategory "rental systems" is on one hand rather widespread such as in the case of international car rental compaenies, and on the other hand it is very local and closed, such as the case for municipal libraries. It may be helpful, especially for libraries, if generic international standards were available, although in this case national standards are often sufficient for local and closed applications.

Table 3.4 Access control systems, personal tracking, rental systems overview

Special application standards related to one subcategory I – International Standards E – European Standards N – National Standards	Access control systems	Personal tracking	Rental systems
Library: implementation of data model for libraries			I

3.4.5 General Assessment of Current RFID Application Standards

As shown previously in the detailed analysis of the different RFID application areas and the application related recommendations, we can see that the area of specific RFID application standards is very diverse and thus generic recommendations on the standardisation for applications is challenging and complex.

By using a SWOT analysis (Strengths, Weaknesses, Opportunities and Threats) there is the possibility to summarise and to extract the most important conclusions of the analysis of the currently existing RFID application standards:

Table 3.5 SWOT analysis of current RFID application standards

Strengths	Weaknesses
International applications specific standards promote RFID technology in those fields where generic standards are not sufficient	Not all applications that need specific standards are yet covered
Stakeholders of different industry sectors (e.g. aviation, automotive, retail, or textile goods) are already aware of the benefits of international application standards	Many existing application standards are only available as national standards
	Many existing application standards have a very narrow focus and cover too specific issues instead of establishing standards for similar applications

Opportunities	Threats
When creating new standards similar applications should be covered by similar international standards as well	Setting up application-specific standards where generic standards would be sufficient
The users' acceptance of application standards will be increased if international application standards are available because supply chains are international	International and national standardisation processes will consume a lot of time and costs
	The acceptance of standards is an important yet also a very difficult issue in general
Application specific standards can support the development in certain areas, e.g. standards for integrated sensors on RFID tags would promote the use of RFID and sensors	A lack of international coordination of setting up application standards could lead to the existence of many similar application standards competing with each other
	Over- and under-specification may occur in certain applications areas
	Where applications are too specific for open standards, proprietary standards might be created. This might result in misleading standards

3.4.6 General Recommendations on RFID Application Standards

In general, broad and generic standards are the best way. However, sometimes applications are very specific, and application-specific standards are needed. The important thing then is to harmonise these application-specific standards worldwide. The following recommendations (and also the recommendations in the above sections) are addressed to the standardisation organisations (i.e. official and industrial/user organisations) as well as to the responsible national and international legislative bodies to pass the necessary mandates for the related official standardisation bodies.

> **Recommendation:** Harmonised international standards and application specific standards should be only established if generic standards are not sufficient

The analysis of the application areas and their subcategories shows that there is a strong need for more harmonised international standards. The need for such a standardisation activity is driven by the increasing level of international commerce in combination with the more international flow of goods. In many application areas where the national standards are not enough there is the need for new international standards based on existing and already established national standards like in the area of product safety and quality.

> **Recommendation:** International cooperation of RFID standardisation organisations

It is important to foster the international cooperation of RFID standardisation organisations like the national legal organisations, international legal organisations, industrial/user organisations, etc. They should cooperate and align their activities in standardisation to prevent the establishment of competitive standards. The existence of competitive standards will increase the uncertainty of the users of RFID technology and will weaken RFID standardisation and RFID technology in general.

3.5 Need for Standards for RFID Sensor Tags

The different stakeholders of RFID technology have identified that additional types of RFID standards will be necessary. For many new RFID applications there is a strong need for the combination of RFID technology together with sensor technology. To push such new applications and the needed mixture of technologies the need of additional standards was identified. Associated with the application of RFID tags with sensor functionality the need for related special packaging standards has also been identified.

The development of such special standards will help to push the emergence of new RFID applications as well as push the related technologies which are necessary to enable these applications.

A newly identified and important issue is the availability of RFID tags, which have integrated sensor functionality. In various businesses such as retail, pharmaceuticals, aviation, automotive, etc. there is a need for integrated sensor functionality in RFID tags. For many applications like for retail pallets with vegetables, it will be helpful to have a very simple RFID-based sensor functionality to measure the current temperature. There is no need for data storage and data processing for many of these simple applications. In some special cases like valuable pharmaceutical products, such as for serums or for system health monitoring in aviation, it will be very useful to have advanced sensor functionality with integrated data storage and perhaps integrated data processing.

Initially only the initiative of IEEE (Institute of Electrical and Electronics Engineers) was concerned with starting the development of standardised sensor interfaces for RFID tags combined with sensors. In the meantime ISO and EPC have also started to work on this topic.

There is the need for RFID sensor standards to push the technological development in order to get a standardised infrastructure i.e. to get standardised interfaces for sensor data, data transfer, etc. for the RFID sensor products. A standardisation of this type will reduce the economic risk for the involved technology developing companies.

For RFID applications such as the afore mentioned RFID tags with integrated sensors, and for many established RFID applications in harsh environments like logistics or industrial manufacturing processes, it is necessary to have RFID tags with special housings, which are resistant against temperature, mechanical impact, corrosion, etc. Currently, only the standard inlay-based RFID tags are mass produced, and can be produced cost-efficiently. If an RFID tag for use in a harsh environment is required, such as for mounting on pallets for logistics within industrial areas, the price for the tags increases drastically.

For this, the EC should support technological development by pushing the establishment of standards for the application of RFID tags in harsh environments to a technical performance standard for harsh environments. This will also reduce the economical risk for developers, and will make this kind of RFID technology and the associated applications a low cost mass product.

3.6 Privacy and Security Standards

3.6.1 *Privacy*

Privacy and data protection is an important and frequently discussed topic in the field of RFID applications. Several standards and directives already exist to protect the privacy of the user and the general public that might be affected by this technology. The European Parliament and the European Council have issued a number of legally binding Directives which must be put into law by the Member States (see Chap. 5.1 for more details):

- 95/46/EC – on the protection of individuals with regard to the processing of personal data and on the free movement of such data,
- 2002/58/EC – directive about the processing of personal data and the protection of privacy in the electronic communications sector (directive on privacy and electronic communications).

To inform the public about the use of RFID in products, all RFID tagged goods shall be marked with an emblem. AIM and EPCglobal have already developed a guideline and an emblem to provide full transparency about tagged goods to the user even if the tagged goods cannot reasonably be used to track the customer (see Chap. 5.1.3.1). No privacy standards have been developed by ISO so far. There is also a need for an ISO standard which forces tagged goods to be marked.

3.6.2 Security

Another important fact is the topic of data security. The use of RFID systems in industry and development allows an easy tracking and tracing of goods and an improvement of manufacturing procedures. But unfortunately also several security risks exist, as shown in Fig. 3.3.

Currently, only a few technology related RFID security standards exist. One of them is the ICAO standard ICAO Doc9303 MRTD, which describes the require-

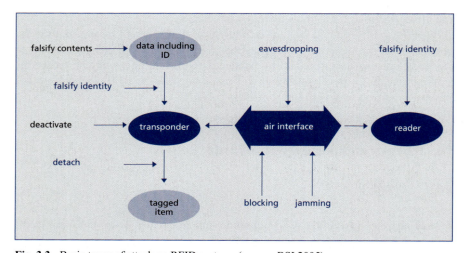

Fig. 3.3 Basic types of attack on RFID systems (source: BSI 2005)

ments and security measures for e-passports based on ISO 14443. Compared to the already used ICAO-standard Machine Readable Passport (MRP), the e-Passport contains additional security features and secure biometric ID confirmation, providing improved border control. The contactless integrated circuit chip embedded inside the e-Passport is in accordance with the standards proposed in the technical report "NTWG-Biometrics Deployment".

Through the fact that there are only a few and very specific standards available, it is not surprising that there is a lot of activity to improve this situation. The document MEMO/07/159 – privacy enhancing technologies, which was published in form of a press release by the European Commission, describes several technologies to increase the security of privacy and data. Technologies that can be used for RFID are the so-called encryption tools. Encryption tools prevent hacking when the information is transmitted over the Internet and support the data controller's obligation to take appropriate measures to protect personal data against unlawful processing.

3.6.3 *Data Security Measures in Air Interface Standards*

Several air interface standards are still offering possibilities to increase data security. The following table gives a short overview which possibilities are supported or can be integrated in transponders using the corresponding air interface standard. The features authentication and encryption are not defined in the standards but some of them refer to another standard which describes these methods, e.g. ISO 7816.

- **Memory locking** is a possibility to protect stored data against unauthorised changes like deleting or overwriting the memory. Usually a bit is set, which prevents the change of the memory.
- A **unique identifier** is a unique number of a device. It can be used as copy and clone protection as well as for data encryption and data covering.
- If a device uses a **random identifier**, a new random number is generated after each reset of the system, which replaces the old UID (Unique Identifier). The random UID can also be used for data covering. The random identifier reduces the risk of eavesdropping since one transponder can not be tracked.
- **Data covering** describes several methods to protect data. These methods do not provide full protection, because the encryption only takes place on a low level. The data is not fully encrypted, but scrambled or deferred. Like the random identifier, data covering reduces the risk of eavesdropping of the data content.
- **Authentication** is the act of establishing or confirming of the identity of the devices. The process of authentication can depend upon one or more authentication factors. Possibilities for authentication are described in ISO standard 9798. One method is the so-called "Three Pass Mutual Authentication". During the process of authentication both communication partners check the know-

Table 3.6 Data security measures in interface standards

Standards	Memory locking	Unique identifier	Random identifier	Data covering	Authentication	Encryption
ISO 14443		X	X		X	X
ISO 15693	X	X				
ISO 18000-6c/ EPC Gen2	X		X	X		
EPC HF	X		X	X		

ledge of a secret cryptographic key. All devices operating in the same application know the same cryptographic key.
- In cryptography, **encryption** is the process of transforming information (referred to as plaintext) so to make it unreadable to anyone except those possessing special knowledge, usually referred to as a key. The result of the process is encrypted information.

Transponders which are conforming to ISO standard 14443 can either have implemented a unique or a random identifier. Furthermore, these transponders offer the possibility to use methods for authentication and encryption in accordance with ISO 7816. In this case the tag is compliant to ISO 14443-4.

ISO 15693 compliant tags offer the possibility to lock memory blocks. This feature is mandatory for tags which are compliant to this standard. A unique identifier is implemented within ISO 15693 tags.

ISO standard 18000-6c is in conformance with EPC Class1 Gen2 Standard. Both standards support the memory locking functionality and provide a random identifier. Further possibilities for data covering are provided.

The EPC HF Standard was developed in dependence on the EPC Gen2 UHF Standard. Therefore, the standard has implemented the same security features like ISO 18000-6C and EPC Gen2 UHF. These features are memory locking, a random identifier and data covering.

Only if data is protected in a certain way, manipulation and system errors can be avoided. The success of RFID technology will be dependent on the fact how good possible implementing of suitable data protection and security methods will work. It is necessary to develop application specific data protection and data security standards which consider the required protection level as well as the capability of the technology.

3.6.4 Recommendations on Privacy and Data Security

> **Recommendation:** Identify the needs for data protection and data security of different application fields and develop corresponding guidelines and suitable security standards

3.6 Privacy and Security Standards

RFID is used in a variety of applications. Passports are equipped with RFID and RFID tags control the production of cars within automotive factories. These different application fields have different needs for data protection and data security. The requirements of specific application fields regarding data protection and data security should be identified and corresponding guidelines should be developed. Further, there is a need for the development of suitable security protocols and algorithms, which are specific for certain applications and which may also contribute to a better protection of personal data. Different application areas need different levels of security. Two options are possible: Either existing standards from other technologies like smart cards can be transferred to RFID or new technologies regarding security standards are developed. Thereby the performance of different tags has to be taken into account. Cheap and easily constructed tags are able to fulfil only a minimum level of security.

Chapter 4
Implementation and Application Guidelines

Although RFID is no longer a new technology, and a basic framework concerning standards has been established, its implementation and widespread use are still in their early stages. It is therefore necessary to provide guidance on technology implementation to companies or pubic entities who wish to use RFID in their business processes. The development of guidelines is no easy task as numerous information demands have to be met. As RFID implementation is currently progressing at a moderate speed, it seems feasible to pose the question whether the guidelines developed so far really serve their purpose in facilitating RFID deployment.

A study by Fraunhofer IPT (Fraunhofer Institute for Production Technology) surveyed around 100 companies with regards to RFID (Fraunhofer IPT 2008). From this study, many surprising results were found. Around a quarter of the firms which were interviewed had already decided in favour of RFID systems without analysing the potential economic benefits through a business case. Thirty percent of the companies also admitted to not having carried out a feasibility study prior to RFID projects. Strüker et al. (2008) interviewed 283 German companies regarding their experiences when implementing RFID and found that those companies that have analysed the benefit of implementing an RFID system for their applications beforehand were significantly more successful. These studies reinforce the assumption that visible and useful guidelines are needed to assist companies who wish to implement RFID systems in their organisations ensuring that all relevant topics will be regarded.

This chapter therefore aims at assessing existing guidelines on RFID, identifying possible information deficits, and developing recommendations on how to improve RFID implementation guidance. In order to analyse guidelines, existing information has been collected via the Internet and desk research. Documents collected consisted of cost/benefit analyses, business models, case studies and other documents regarding experiences with RFID implementation.

4.1 Requirements of Guidelines

In order to analyse RFID guidelines, a framework for analysis has to be established. For this purpose, we are proposing a two-dimensional approach to include the two most important dimensions that have to be addressed to create a maximum of practicability of guidelines in daily business processes.

4.1.1 The RFID Reference Model

The first dimension is application driven. The RFID Reference Model has been introduced in Chap. 2.1 as an attempt to structure the different application fields in order to be able to evaluate requirements in terms of standards, technological components, and data protection for individual applications. We also used it as format to evaluate RFID guidelines regarding the scope of guideline coverage across the different RFID application fields.

4.1.2 The RFID Implementation Checklist

The second dimension is the addressee approach. The RFID Stakeholder Model is depicted in Chap. 2.2, giving a structured overview of different groups that might be directly or indirectly involved in RFID technology, or perhaps simply interested in the topic. The second dimension we used for our analyses can be derived from the RFID Stakeholder Model. Different stakeholders, e.g. groups within companies, as well as different stakeholders outside of companies have different information requirements. These different requirements should be taken into account when drafting guidelines.

One key result of the expert discussions and the research conducted in the field of RFID guidelines is that:

- most guidelines which are available are too broad in focus in order for them to be useful for a company or institution wishing to employ the technology, and
- it is crucial to clarify the individual requirements that different addressees expect to be fulfilled by a certain guideline.

As a way of giving the necessary guidance, an extensive RFID Implementation Checklist has been established in the course of the work summarising guideline requirements. This checklist has been developed and discussed in-depth with various RFID experts and it serves a twofold purpose; firstly it is the framework for the analysis within this chapter, and secondly it can be used as guidance for the establishment of application specific guidelines at a company or entity level in the future.

4.1 Requirements of Guidelines

There are already several RFID checklists available to potential RFID users; for a brief insight see Sieker, Ladkin and Hennig (2005), Sicher im Netz e.V. (2008), or Netzwerk elektronischer Geschäftsverkehr (2008).

Most of them, however, focus on only one or two topics (usually privacy) and are therefore too narrow to give a complete overview of all the aspects that need to be taken into consideration when a company plans to adopt RFID technology. In general, checklists give guidance for newcomers to RFID technology. They provide an easy overview and can serve as the basis for drafting guidelines for a wide range of individual applications. They are more flexible and adaptable than fixed guidelines and therefore do not run the risk of becoming outdated due to the rapid technological progress in the field of RFID. Elaboration and updating of the RFID Implementation Checklist can be an opportunity to foster and enhance the dialogue between RFID users and stakeholders.

Challenges with respect to checklists are to make sure that the responsibilities for regular updates are clear, and to stress the fact that the checklist alone does not deliver a fixed solution for an RFID implementation problem. This solution can only be reached by taking the checklist as a basis for an inter-company process involving all relevant parties with regard to the planned RFID adoption.

Several addressees have been included in this checklist. These addressees have specific information requirements, depending upon their specific position.

- **Decision Makers** are in charge of strategic decisions and are usually not deeply concerned with the technical details of the system plan, but with the overall effect it will have on the business, such as cost/benefit analyses and ROI, organisational fit, and potential competitive advantages.
- The **IT Department** group is concerned with technical information regarding the physical integration of the RFID system into the current IT infrastructure. This includes information such as the HW/SW structure, security issues and standards.
- The **Process Management** group is interested in a potential change of processes when implementing an RFID system, the integration of new processes into the existing business processes, and evaluating how processes can operate more efficiently.
- The **Legal Department** handles the legal issues associated with RFID and requires information such as related legislation, to ensure that the firm is complying with all relevant national and EU legislation.
- The **Human Resources (HR) Department** handles personnel issues; therefore it would need to be provided with information regarding topics such as health implications from the use of RFID or also employee rights.
- The **Communication Department** requires information on RFID so that it can communicate information to various parties. This could include themes such as providing information about how privacy is ensured or how the technology works to consumers to generate trust.

The following table shows the RFID Implementation Checklist, stating the different addressees and their specific guideline requirements.

Table 4.1 RFID Implementation Checklist

Addressee of guideline	Specific information requirement
Decision makers	New business opportunities
	Organisational fit
	Case studies as a benchmark
	Data for calculating a business case
	Cost-benefit analysis
	Time frame
	Resources needed for implementation
	Risk assessment
	Measurability of economic success
	Non-monetary positive effects
	Legal aspects
	Overview of current RFID technology
	IT security
	Technological requirements
	Alternate technologies
	Outlook, future developments, roadmap
	Data sharing systems
IT department	Standards
	Data storage
	Tag technology
	IT security
	Reliability
	HW/SW architecture
	Networking
	Frequency restrictions & international interoperability
	Interferences
	Interfaces
	Co-existence/Migration Classical Auto ID to RFID
	Deactivation
Process management	Information about reference processes to assess fit of guideline to own application scenarios
	Attachment of tags to item
	Awareness of technical properties of different RFID techniques
	Co-existence/Migration Classical Auto ID to RFID
	Environmental conditions
	Standards
	Known fields for potential improvements of processes
	Outlook, future developments, roadmap
Legal department	Ensuring compliance with relevant legislation and existing self regulations
	Guidance when written consent for data collection is necessary
	Depicting legislative process concerning data protection
	Explaining legitimate purpose for data collection
HR department	Overview of current RFID technology
	Scope and purpose of data collection
	Health issues
	Compliance with employee rights
	Influence on employment
	Training for employees
Communication department	Information for stakeholders
	Communicating security measures to generate trust
	Strategic communication campaign

This list does not target specific RFID applications; it can be used by companies or organisations as guidance when implementing an RFID system and to establish detailed guidelines that are tailored to the companies' or organisations' needs. Depending on the specific application area, some of the topics might be more or less important for the addressees in question.

Therefore, the RFID Implementation Checklist serves two purposes – one general purpose and one within the context of this study. As explained above, the RFID Implementation Checklist serves as guidance for companies and entities that are planning to adopt RFID technology. For the content analysis of guidelines conducted in this study, the RFID Implementation Checklist delivers the compilation of addressee-specific requirements – one cornerstone in evaluating whether a guideline really fulfils the needs of those who are expected or intended to use it. The RFID Implementation Checklist can be downloaded from the CE RFID website (www.rfid-in-action.eu).

With the application and the addressee approach, a comprehensive framework for the following analysis has been established. The framework takes into account that guidelines need to be application-specific to be most useful, and that different addressee groups need different information in order to be able to implement RFID systems in their company or entity.

4.2 Analysis of Existing Guidelines

Basic research on RFID guidelines has been conducted via desk research and expert interviews. Only publicly available guidelines have been included in the study as it gives an impression of how interested parties would research the information they would need if they were in the initial planning process for an RFID implementation project. A preliminary working definition of "RFID guideline" has been established as "any information document providing information about setting up an RFID system".

In order to establish the research project in the broadest way possible, the approach, analysis, and the results have been repeatedly discussed with stakeholders including technology suppliers, users, their business associations, and also standardisation organisations and research institutions.

4.2.1 Method

To analyse and categorise the guidelines found, a content analysis was conducted. The content analysis provides a good basis to comment on guidelines in a structured and standardised way, rather than writing a comment for each guideline. The results allow statements and conclusions to be made on guidelines in a more analytical manner.

Content analysis is a standard methodology in social sciences for studying the content of communication. Content analysis as an empiric method of data collection can be applied to all kinds of recorded communication, and therefore can also be used in the analysis of "guidelines" as written textual documents. It enables the researcher to include a large amount of information and to systematically identify its properties by detecting the important structures of its communication content (Weber 1990, Shapiro and Markoff 1997).

The advantage is that qualitative data can be transferred into quantitative data using systematic and replicable techniques to allow a structured analysis using statistical methods. Deffner (1986) classified content analysis methods into three broad types: human scored schema, individual-word-count systems (usually computerised), and computerised systems using artificial intelligence. For analysing guidelines the human scored schema will be used to evaluate the topics of each guideline. Quantitative analytical steps will be performed too, as the value of a guideline for a user depends on the level of information provided (e.g. the number of pages of a guideline can be a first, however weak, indicator of the depth of the information provided).

Initially the unit of analysis has to be defined. After that, code units and the respective coding rules have to be set up (Morris 1994). The unit of analysis describes the communication content to be analysed, in this case an RFID guideline. The code units

Table 4.2 Example of code book

Addressee code	Addressee	Requirement code	Requirement	Description
1100	Decision makers	1101	New business opportunities	Depicted potential business opportunities that can emerge when implementing RFID
		1102	Organisational fit	Discussion of impacts on the business process
		1103	Case studies as a benchmark	Case studies that can serve as a benchmark
		1104	Data for calculating a business case	Clear outline of which data is needed to calculate the own specific business case
		1105	Cost and benefit analysis	Detailed and clear cost and benefit analysis so that implications for the own company can be derived
		1106	Time frame	Description of how long the implementation of RFID technology may take
		1107	Resources needed for implementation	Resources (financial and human resources) which are needed to implement the system

can be formal, e.g. number of pages, author, year of publication, or contextual, depending on the research interest and needs to be defined in detail. The code book describes the procedures for analysis and supports the coder with coding advice and rules. A precondition for the code book is a clear structure, so that it can be applied even by persons who are not involved in the topic. The code book also includes the coding sheet set up according to the content of the codebook and facilitating the coding procedure. Table 4.2 shows an extract of the coding sheet used for the analysis.

Reliability and validity are important quantity measures for a content analysis. In order to make valid inferences from the text, it is important that the classification procedure is reliable in the sense of being consistent; different people should code the same text in the same way (D'Aveni and MacMillan 1990). To ensure an objective analysis the coding results and the resulting report has also been sent to the writers of the analysed guidelines to ensure that they agree with the analysis of their respective guidelines.

Assuming that guidelines are written for certain addressees, a codebook was prepared allowing the analysis of the guidelines found with regard to, for instance, fulfilment of requirements of different addressees, frequency and depth of certain topics or communication strategies for different stakeholders. Table 4.2 shows an extract of the developed code book, in particular the part of the decision makers as addressees and their requirements to the content of a guideline. The code book is a detailed and elaborated list describing all requirements of all addressees to make a clear and distinctive statement for the analysis of what is understood. The complete code book can be found in the report of this research project (Gampl et al. 2008a).

4.2.2 Initial Categorisation

Initially RFID documents that could be regarded as guidelines or that have been regarded as guidelines by their respective authors were collected. 71 RFID documents were identified until October 2007 by conducting research via the Internet. The search phrases used to find RFID documents were for example "RFID" in combination with "guide", "guidebook", "handbook", "guideline", or "guidance". The list compiled has been reviewed by experts from industry and academia, e.g. CE RFID members and additional contributors. It can therefore be assumed that it contains most of the relevant public RFID guidelines. The result contains textual documents, as well as calculators such as Excel based tools or programs, for evaluating RFID business cases. The documents have been categorised into the following groups:

- Generic information on RFID systems (23)
- Privacy (21)
- Communication strategies to inform consumers/stakeholders (3)
- System implementation (6)
- SMEs and RFID (5)
- Special business applications (8)
- Costs and benefits of an RFID implementation (5)

Most documents belong to the first two groups (44 out of 71), i.e., "generic information on RFID systems" and "privacy". It was difficult to find useful case studies and cost-benefit analyses, as they were usually too generic to support companies to draw conclusions for their own application scenarios. For a list of all 71 documents found please refer to the extended report (Gampl et al. 2008a).

By looking at these 71 documents, the high variation regarding their content and their level of detail becomes apparent. Not all of the documents seem to match the criteria for a guideline. As next step, the working definition "any information document providing information about setting up an RFID system" was further elaborated and the definition of the term "RFID guideline" for this research project has been established.

4.2.3 Definition of RFID Guidelines

One of the central results was that current and future users of RFID applications require application-specific guidelines. Requirements and necessary specifications differ greatly between different RFID applications and cannot be covered in one single unifying document. This view is also supported by the "European Policy Outlook RFID" which was elaborated during the German EU Presidency by stakeholders from industry as well as consumer and data protection organisations, representatives from the European Commission and government bodies. It is stated with regard to the issue of privacy that it was one of the challenges of the RFID debate "to clearly differentiate between different application fields, making sure that the use of the technology is not compromised where privacy is not an issue and clearly highlighting problems where privacy, freedom from paternalism, or competition are at stake" (BMWi 2007). Keeping this in mind, for the context of this study the following definition for the term "guideline" has been established:

> "Guidelines are domain-specific documents assisting companies and organisations when implementing and using RFID systems with regard to possibly affected stakeholders. Such guidelines need to address specific concerns of entities within companies and organisations that implement or use RFID systems."

The term "domain" in this respect refers to the RFID Reference Model with its eight different application domains. "Addressees" of guidelines are "RFID users", i.e., companies or entities within these companies that use or plan to use RFID technology in their specific processes. The term "stakeholder" describes all interested parties, possibly also outside the company, which might be interested or affected by a company's introduction of RFID. By stating that a guideline needs to address "specific concerns of entities" it suggests the idea that even a domain specific approach may not be detailed enough for a useful guideline. According to the RFID Reference Model an RFID implementation guideline has to be written for a specific subcategory, e.g. within the broad application field Logistical Tracking & Tracing it has to address either topics of the subcategory In-House Logistics, Closed-Loop Logistics or Open Logistics.

4.2.4 Process of Analysis

The first step of evaluating a document is to assess whether it is domain-specific or not. Following the RFID Reference Model, "domain-specific" implies that a document specifies concerns of certain RFID application fields, e.g. "Logistical Tracking & Tracing" or "Product Safety, Quality and Information". "Application-specific" implies that it specifies concerns of single applications within a domain.

If a document is not domain-specific, it can be treated as a general information document about RFID technology. If it is domain-specific, the document needs to be further analysed to see if it is also application-specific. This step is necessary, since according to the definition, guidelines "need to address specific concerns of entities within companies and organisations that implement or use RFID systems". These specific concerns can only be addressed if the guideline is application-specific. If it is not application-specific it can also be treated as a general information document about RFID technology. If the analysed document fulfils both requirements, being domain and application-specific, it is re-

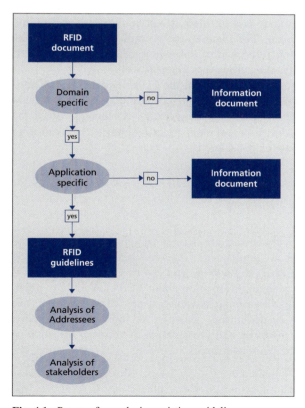

Fig. 4.1 Process for analysing existing guidelines

Fig. 4.2 Process diagram for analysis of requirements of addressees

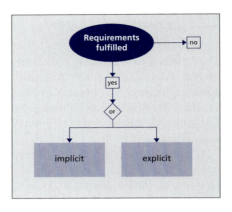

garded as a guideline and will be further evaluated with respect to the addressees and possibly mentioned stakeholders. The process diagram is depicted in Fig. 4.1.

The analysis of the addressees will be conducted according to Fig. 4.2. In this step the specific requirements of the addressees will be taken into account. For each of these requirements the question will be answered whether the single requirement is fulfilled or not. In case it is fulfilled, it will be further specified into "explicit" or "implicit" fulfilment. The requirement is implicitly fulfilled if the topic is only named. An explicit fulfilment of the requirement means that a certain topic is described in detail and examples are given as well.

The process described was applied to 67 out of the 71 RFID documents identified. The omitted four "documents" are cost/benefit calculators. The content analysis to evaluate guidelines aims at analysing textual documents. Since the calculators are, for example, Excel-based tools or programmes for evaluating RFID business cases, they cannot be considered.

4.2.5 List of Guidelines Analysed

As a result, 20 out of these 67 documents are defined as application-specific and could therefore be categorised as "guidelines". The remaining 47 documents do not fulfil the requirements for being a guideline and can be categorised as information documents, delivering generic information about RFID technology, application examples, and generally describing advantages and disadvantages. The following table depicts these 20 documents, showing an internal identification number (also used for further analysis and presentation of results), the authors, a short description for each, the main addressees according to our evaluation, the domain and the publishing date.

4.2 Analysis of Existing Guidelines 57

Table 4.3 List and description of RFID guidelines analysed

#	Title	Editor/Author	Description and Main Addressees	Domain	Year
70	Radio Frequency Identification (RFID) in Retail Consumer Privacy Code of Practice	Australian Retailers Association	This document contains an industry code of practice regarding consumer privacy in the retail sector. The guideline specifies how retailers can inform consumers about RFID. It suggests how personal data should be handled when using RFID, how staff can be trained on the use of RFID and consumer rights, and how to handle complaints regarding RFID. Due to the communication themes, the addressees of this guideline are mainly communication parties. The guideline is also legally orientated, as it discusses privacy issues. Main addressees: Communication Department, Legal Department, HR Department. (20 pages)	Product Safety, Quality & Information	2007
22	RFID-Studie 2007: Technologieintegrierte Datensicherheit bei RFID-Systemen (RFID Study 2007: Technology integrated data integrity in RFID systems)	BMBF (Bundesministerium für Bildung und Forschung)	This guideline defines detailed methods for a secure application by using examples of there concrete application scenarios (Automotive production, retail and the pharmaceutical supply chain). The overall study also gives recommendations for further research and technology development. Main addressees: Decision Makers, IT Department, Process Management (156 pages)	Production, Monitoring & Maintenance Product Safety, Quality & Information Logistical Tracking & Tracing	2007
52	Using HIBC Standards with RFID: An Implementation Guideline	HIBCC (Health Industry Business Communications Council)	This guideline has been written as a technical advisory for coding the HIBC data structure in the healthcare sector. It also details how the HIBC Supplier Labelling Standard can operate alongside other standards for the labelling of goods and assets. Main addressees: IT Department, Process Management, Decision Makers (30 pages)	Logistical Tracking & Tracing	2007

Table 4.3 *(continued)*

#	Title	Editor/Author	Description and Main Addressees	Domain	Year
71	The Pharma Guide to RFID	UPM Raflatac	This guideline was developed for pharmaceutical companies who are interested in implementing RFID in their organisation. The guide gives an overview of RFID technology, information on creating a business case, and case studies from early adopters. Information related to process management has been included here, referring to how RFID can operate and enhance existing processes in an organisation. IT department related information has been included in the form of providing tag and frequency specifications. Information targeted at decision makers has also been added, as the guideline includes successful case studies and outlines business benefits from RFID implementation. Main addressees: Process Management, IT Department, Decision Makers (35 pages)	eHealth Care Product Safety, Quality & Information	2007
64	RFID for the Healthcare Sector	Informationsforum RFID, Germany	This document details the potential breadth of application for RFID technology in the healthcare sector. It contains detailed information on various case studies which have been used RFID in the healthcare sector. The case studies included in this document ensure that a broad range of addressees are addressed. Main addressees: Communication Department, Process Management, HR Department (20 pages)	eHealth Care	2007
58	RFID-Leitfaden für die Automotive-Branche (RFID Guideline for the Automotive Sector)	RFID Support Center	This guideline highlights important technical information and application examples of the use of RFID in the automotive sector. It provides two use cases when RFID has been used in the automotive sector, and advice from companies who have already implemented RFID. The process and IT departments are included here as many of the case studies talk about RFID and process integration. The decision makers are addressed in this guideline as they give real life examples for decision makers to see how RFID can benefit business processes. Main addressees: Process Management, Decision Makers, IT Department (31 pages)	Production, Monitoring & Maintenance Logistical Tracking & Tracing	2007

4.2 Analysis of Existing Guidelines 59

#	Title	Editor/Author	Description and Main Addressees	Domain	Year
56	Einführung von RFID zur Optimierung der Produktionsplanung und -steuerung (Introduction of RFID to optimize production planning and control)	B. Scholz-Reiter, C. Gorldt, U. Hinrichs, J. T. Tervo	This guideline contains information describing key aspects of RFID implementation in SMEs to optimise production planning and control. It gives a rough overview on the steps that should be taken to first of all evaluate if RFID is the right solution for the organisation, and as secondly what has to be considered when implementing an RFID system. The guideline addresses mainly technical topics; however issues relevant for the Human Resources department have also been included. Main addressees: Process Management, Decision Makers, IT Department (5 pages)	Production, Monitoring & Maintenance	2007
47	RFID Guidelines and Requirements	Wal-Mart	This document summarises the main standards and requirements for the Wal-Mart RFID Roll-out. The document outlines a deployment plan, process guidelines and requirements, data exchange guidelines and requirements, and technological information. The process department has been addressed as a section detailing process guidelines and requirements has been included. The IT department has also been addressed, as technology requirements and data exchange standards have been outlined. Decision makers are also addressed in this document, as a deployment scope for the roll-out has been included. Main addressees: Process Management, Decision Makers, IT Department (25 pages)	Product Safety, Quality & Information Logistical Tracking & Tracing	2007
10	Social and labour implications of the increased use of advanced retail technologies	International Labour Organization (ILO, United Nations, Geneva)	This guideline provides an overview of the potential societal and labour implications which may arise due to the increased use of advanced retail technologies, specifically RFID technology. It includes an overview of the retail sector globally, the impacts that RFID could have on the retail market, and also the impacts that RFID will have on society and the labour market. The guideline has a broad scope of addressees as it analyses which areas RFID could be implemented into, in several broad scenarios. Main addressees: Process Management, HR Department, Decision Makers (63 pages)	Product Safety, Quality & Information Logistical Tracking & Tracing	2006

Table 4.3 (continued)

#	Title	Editor/Author	Description and Main Addressees	Domain	Year
32	Practical Tips for Implementing RFID Privacy Guidelines	Information and Privacy Commissioner (Ontario, Canada)	This guideline allows companies to address privacy issues associated with RFID implementation, specifically how to comply with relevant legislation and best practices. The guideline also discusses methods of how to create individual company policies on RFID, and how to communicate information on RFID to consumers. The guideline addresses mainly the communication and the legal department, as the guideline is about ensuring consumer privacy, and communicating with the consumer regarding privacy issues related with RFID. Main addressees: Communication Department, Legal Department (3 pages)	Product Safety, Quality & Information	2006
50	Guidelines for the METRO Group RFID Roll-out	METRO Group	This guideline was written by METRO Group for partners who are part of the RFID roll-out. The document explains why RFID is important for supply chain management (SCM) and how it can benefit businesses. It outlines the prerequisites for the adoption of the technology, and technical requirements for the programme. Further support options from the METRO Group are also detailed here. The guideline addresses the various addressees, as the roll-out would affect all of these groups. Main addressees: Communication Department, Process Management, IT Department (44 pages)	Product Safety, Quality & Information Logistical Tracking & Tracing	2006
66	Shrinking the Supply Chain Expands the Return: The ROI of RFID in the Supply Chain	Thomas Pisello, An Alinean White Paper	The White paper uses three case studies to present examples for return on investment (ROI) gained from RFID applications in the fields of Manufacturing, Warehouse and Distribution Solutions, and Retail. Main addressees: Decision Makers, Process Management, IT Department (16 pages)	Production, Monitoring & Maintenance Logistical Tracking & Tracing Product Safety, Quality & Information	2006

4.2 Analysis of Existing Guidelines

#	Title	Editor/Author	Description and Main Addressees	Domain	Year
51	A Guideline to RFID application in supply chains	REGINS-RFID	This guideline describes RFID technology in detail, and describes how RFID can be implemented into supply chains. The guideline is divided into three main areas – RFID technology and related applications; the role of RFID in supply chains; a roadmap for RFID implementation. The addressees are addressed on quite a high level, due to the fact that this guideline has a broad overview of RFID and related subjects. A very brief roadmap has been included. Main addressees: Communication Department, Process Management, IT Department (108 pages)	Logistical Tracking & Tracing	2006
45	Guidelines on EPC for Consumer Products	EPCglobal	This is a short guideline concerning EPC and consumer products. The guideline gives an overview of RFID, an outline of the benefits, and the privacy implications which come with RFID. From these privacy implications, EPCglobal then outlines how consumers can be informed about RFID, and how EPCglobal informs customers about the presence of RFID. The guideline only addresses the communication and the legal department as the guideline is only about privacy and communicating information to consumers. Main addressees: Communication Department, Legal Department (2 pages)	Product Safety, Quality & Information	2005
33	EPCglobal Guidelines at the METRO Group	METRO Group	This is a 1 page picture based guideline which details the use of EPCglobal guidelines in the METRO organisation. It explains where consumers can find information related to the use of RFID, the obligations to inform consumers when RFID is being used, the deactivation opportunities, and also the related privacy issues. As this guideline is aimed at consumers, and handles legal issues associated with RFID, it is clear that the 2 main addressee groups are the communication department and the legal department. Main addressees: Communication Department, Legal Department (1 page)	Product Safety, Quality & Information	2005

Table 4.3 (continued)

#	Title	Editor/Author	Description and Main Addressees	Domain	Year
61	A UK Code of Practice for the use of Radio Frequency Identification in Logistics and Supply Chain	National RFID Centre (UK)	This guideline is a template for a UK code of practice for the use of RFID in logistics and supply chain, specifically privacy and safety. The communication department is an addressee as the guideline contains information on how to inform consumers about the use of RFID. The legal department is addressed here due to the fact that consumer privacy issues are involved. Decision makers are also addressed here as RFID benefits are detailed in the document. Main addressees: Communication Department, Legal Department (1 page)	Logistical Tracking & Tracing	2005
62	A UK Code of Practice for the use of Radio Frequency Identification in Manufacturing and Warehouse	National RFID Centre (UK)	This guideline is a template for a UK code of practice for the use of RFID in production, specifically privacy and safety. The communication department is an addressee as the guideline contains information on how to inform consumers about the use of RFID. The legal department is addressed here due to the fact that consumer privacy issues are involved. Decision makers are also addressed here as RFID benefits are detailed in the document. Main addressees: Communication Department, Legal Department (1 page)	Production, Monitoring & Maintenance	2005
63	A UK Code of Practice for the use of Radio Frequency Identification in Retail Outlets	National RFID Centre (UK)	This guideline is a template for a UK code of practice for the use of RFID in retail environments, specifically privacy and safety from a consumer viewpoint. The communication department is an addressee here as the guideline states that consumers will be informed if RFID is in use. The legal department is addressed here as privacy related issues are explained. The decision makers group is also addressed here, as the benefits of RFID are briefly outlined. Main addressees: Communication Department, Legal Department (1 page)	Product Safety, Quality & Information	2005

4.2 Analysis of Existing Guidelines

#	Title	Editor/Author	Description and Main Addressees	Domain	Year
60	Guidance from AIM Global's RFID Expert Group: Proposed guidelines for the use of RFID-enabled labels in military logistics	AIM (Association for Automatic Identification and Mobility)	The guideline details information regarding the use of RFID labels and tags to track goods and assets in a military environment. It contains detailed information on tag placement, specifically how to place the tag on an object and achieve the best signal possible. The guideline also contains information on tag-related standards. According to the technical focus of that guideline the addressees are IT Department, Process Management and Decision Makers. Main addressees: Process Management, IT Department, Decision Makers (38 pages)	Logistical Tracking & Tracing	2004
59	Guidelines for Using RFID Tags in Ontario Public Libraries	Information and Privacy Commissioner (Ontario, Canada)	This guideline describes general implementation procedures for RFID in libraries in Ontario, Canada. The guideline focuses on privacy issues which could become apparent due to the use of RFID. The communication department is the main addressee in this guideline as the content regards communicating privacy issues to consumers. The HR department is addressed due to the fact that a guideline regarding the responsible handling of data is included. The legal department is included as many privacy issues are covered. The process management department is addressed by including guidelines on potential standardisation possibilities between other RFID systems. The IT department is addressed in this guideline, as some technical information regarding RFID tags is included. The decision makers addressee group is included as considerations have been listed which need to be taken into account when implementing an RFID system. Main addressees: Communication Department, Legal Department, Decision Makers (15 pages)	Sports, Leisure & Household	2004

4.3 Quantitative Analysis of Guidelines

Before conducting the content analysis of the 20 RFID guidelines, formal categories for the analysis were defined such as authors, language, number of pages, and year of publication. The following section gives an overview of the formal categorisation.

4.3.1 Formal Categories

From the 20 guidelines analysed, five were written by business organisations, five by companies and five by governmental organisations. NGOs (Non-Governmental Organisations) could twice be identified as authors. The remaining three guidelines came from other sources, i.e. universities and (public) research institutes, and collaborations between research institutes and companies (Table 4.4). Depending on the authors, guidelines might have a specific focus, e.g. companies would probably write more specific guidelines whereas business associations would rather elaborate the topics in a broader and more general sense. This categorisation shows a fairly balanced situation of the sources of the analysed guidelines. As a first conclusion it is evident that the authorship of RFID guidelines is spread quite evenly across different institutions.

14 of the guidelines analysed are written in English. Four guidelines are written in German, and two are available in both languages, German and English. The year of publication was defined as another formal category. Eight guidelines were published in 2007. The publication date for five RFID guidelines was 2006 and it was 2005 for another five. Only two RFID guidelines were published in 2004 (Table 4.5). The majority of the guidelines analysed have been published within the last two years allowing them to give an up to date overview. The higher number of guidelines in the recent years supports the argument of RFID technology being a fast growing and developing technology that is just now receiving increased public attention and is now starting to offer business opportunities that are attractive to an increasing number of companies.

Table 4.4 Overview of authors

Author	Number of Guidelines
Companies	5
Business organisation	5
Governmental organisations	5
NGOs	2
Others	3
Total	20

4.3 Quantitative Analysis of Guidelines

Table 4.5 Year of publication

Year of Publication	Number of Guidelines
2004	2
2005	5
2006	5
2007	8
Total	20

Since all RFID guidelines analysed are textual documents, the quantity of content of these guidelines can be measured considering the number of pages (Fig. 4.3). The value of a guideline for a user depends on the level of information provided, and the number of pages can be seen as a first – albeit, weak – indicator for the level of information provided. Seven guidelines have less than 10 pages. Five guidelines have 11–30 pages and four guidelines are between 31 and 40 pages long. The longest guideline counts 156 pages. The diagram shows the 20 analysed RFID guidelines by their number of pages. The x-axis represents the guidelines by their identification number (see Fig. 4.3), with a decreasing number of pages from left to right. The y-axis represents the number of pages for each guideline. It can clearly be seen that there is a steep decrease in the number of pages with a majority of guidelines being quite short. As guidelines are expected to give very clear and detailed information guidelines with only a few pages seem to be of low value for users. The analysis shows that several guidelines are very short and can therefore not give a complete overview of the topic.

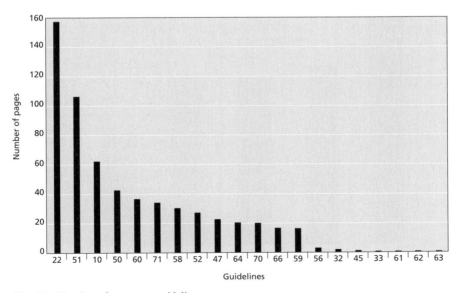

Fig. 4.3 Number of pages per guideline

4.3.2 Addressees of Guidelines

Following the addressee approach explained above, an effective way to evaluate the content of a guideline is to analyse whether the specific requirements of each addressee group have been met and which addressees have been addressed. Figure 4.4 shows the number of guidelines that address each of the different addressee groups (each guideline can address more than one addressee): The most relevant addressee is the group of decision makers that could be identified in 19 out of the 20 guidelines, followed by the IT department with an occurrence in 14 out of 20 guidelines. 13 out of 20 guidelines address each the group of the legal department, the communication department and also the process management. Lower focus is given to the HR department, as only nine out of 20 guidelines target this addressee group.

As a next step, the analysis focuses on the number of addressees addressed in each guideline. Figure 4.5 shows that none of the RFID guidelines analysed has only one addressee. Most guidelines refer to three addressees (eight out of 20 guidelines). Six guidelines address all six groups of addressees.

To summarise, all guidelines have at least two addressees and most of the analysed RFID guidelines (18 out of 20) have three or more addressees.

In order to analyse the qualitative content of an RFID guideline it is necessary to determine whether the specific requirements of the defined addressees are fulfilled. The addressees and their specific requirements were defined prior to the content analysis and are listed in the RFID Implementation Checklist (Chap. 4.1.2).

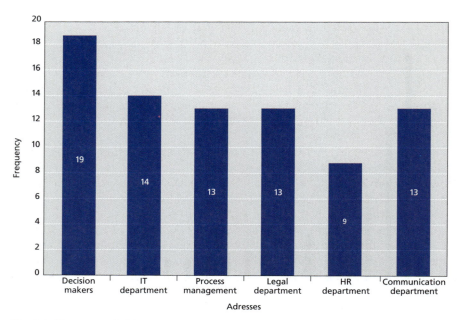

Fig. 4.4 Frequency of addressees

4.3 Quantitative Analysis of Guidelines

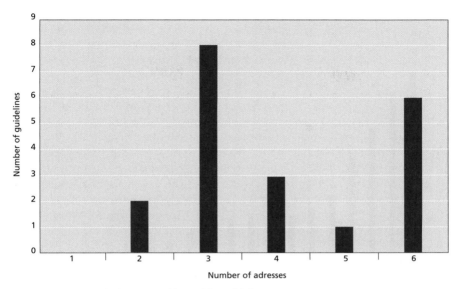

Fig. 4.5 Number of addressees addressed by guidelines

As a coding rule, it was defined that the fulfilment of a requirement can be either explicit or implicit. Initially all implicit and explicit fulfilment has been summed up to "fulfilment". The results can be depicted for all addressees, and also split by each addressee group. Figure 4.6 shows the fulfilment of requirements for all addressees. The x-axis represents the 20 guidelines by their identification number and the y-axis represents the total fulfilment of requirements for each guideline sorted by a decreasing fulfilment of the requirements from left to right.

As an example, the guideline on the left (first column) fulfils approx. 82 % of all requirements set up and the guideline on the right (last column) fulfils about 15 % of all requirements. The median fulfilment of requirements of all guidelines is approx. 31 %. (The median divides a sample into half, meaning that 50 % (ten guidelines) fulfil more than 31 % of all requirements and the other 50 % fulfil less than 31 % of all requirements.)

Because the guidelines are sorted by their fulfilment of requirements this may mislead one to understand Fig. 4.6 as an absolute and objective quality ranking. However, as stated before it is not possible to draw such conclusions concerning the quality of guidelines. It can only be stated how many requirements have been fulfilled; conclusions regarding the respective quality are not consequently valid. A guideline that only fulfils 30 % of all requirements may be perfect for the application that it has been written for and a guideline fulfilling 70 % of all requirements written without an application in mind can be less useful. The analysis and the results give an overview of the guidelines available, describe them and give hints on further need of information or action. For the fulfilment of requirements of all guidelines for all single addressees please refer to the extended report of this research project (Gampl et al. 2008a).

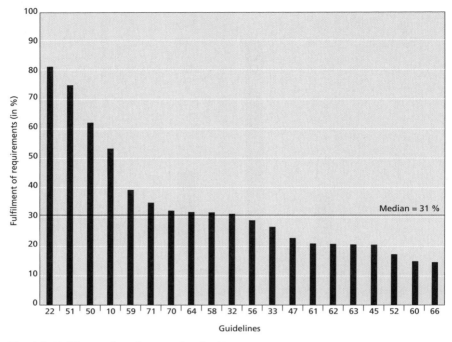

Fig. 4.6 Fulfilment of requirements for all addressees

The analysis of the fulfilment of requirements can be depicted for each addressee using the values minimum, median, and maximum (Fig. 4.7). The x-axis represents the six addressees. The y-axis represents the fulfilment of requirements in percent. The diagram shows the fulfilment of the specific requirements with their maximum and minimum values, and the resulting medians.

The minimum is always 0 %, because for each addressee exists at least one guideline that does not address the requirements of this addressee. In three cases (IT department, process management and communication department) at least one guideline fulfils 100 % of the requirements set up. When looking at the median values it becomes evident that the requirements of Human Resources departments are rarely addressed. For the addressee groups of decision makers and IT department, low values are evident as well. The two addressee groups, process management and legal department show higher values. Communication department shows the highest median fulfilment of guideline requirements.

This result is surprising, as most of the RFID applications up to now are in the field of logistics and production (Strüker et al. 2008) with very little or no interfaces to consumers. We would have expected higher values for the human resources group, as employees will be using RFID systems. However, these results show that RFID is a topic that has recently gained a lot of public attention and it seems that even if consumers are not affected by the implementation of an RFID system, companies are still thinking about how to communicate such a decision.

4.3 Quantitative Analysis of Guidelines

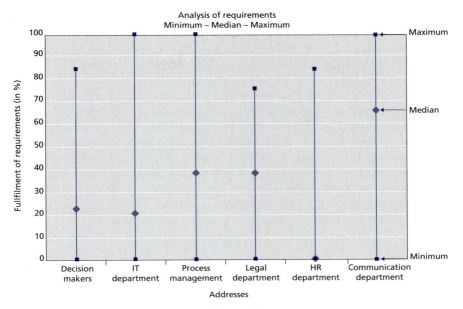

Fig. 4.7 Analysis of requirements – according to addressees

As the number of addressees addressed by each individual guideline differs according to Fig. 4.5, the suggestion is close that guidelines can be either general (addressing most of the addressees) or specialised (addressing only a selection of addressees). Since the analysis of addressees and their requirements has shown a quite low fulfilment of requirements, the question arises as to whether specialised guidelines fulfil more requirements than general guidelines. It has been defined that general guidelines address five to six addressees whereas specialised guidelines address one to four addressees.

Table 4.6 Maximum fulfilment of requirements per addressee

Addressees	Maximum fulfilment (in %)	
	General guidelines (n=7)	Specialised guidelines (n=13)
Decision makers	82	59
IT department	100	67
Process management	100	100
Legal department	75	75
HR department	83	33
Communication department	100	100

Upon evaluation, seven out of 20 guidelines belong to the group of general guidelines, having five to six addressees. 13 out of the 20 analysed RFID guidelines are specialised, having one to four addressees. Table 4.6 shows the differentiation between general and specialised guidelines and the fulfilment of the requirements (in percent) for the respective addressees.

This comparison shows that specialised guidelines do not necessarily meet more requirements than general guidelines. The values for fulfilment of requirements for the general guidelines are always higher or the same as for specialised guidelines. Taking the mean value for this evaluation leads to the same results. This shows that it is possible to address many addressees and still satisfy their informational needs. Writing guidelines only for one specific group to address all its needs does not always seem necessary. As described in the RFID Implementation Checklist, some of the groups of addressees demand similar information. By addressing these groups in one document synergies can be realised.

4.3.3 Consideration of Stakeholders

Following the addressee approach it has been emphasised that addressees do not only act within the RFID user company or entity itself. They reach out beyond the user entity as they interact with groups on the outside. For the further analysis of RFID guidelines we have defined groups of "stakeholders" that could be of interest for the addresses.

Addressees that interact with stakeholders have specific requirements which need to be fulfilled in the guidelines. These respective requirements are related to stakeholders' specific views and perceptions of RFID technology and to its implementation within the user entity. The content analysis considered these aspects in the definition of the addressees' requirements as well. Figure 4.8 shows some of

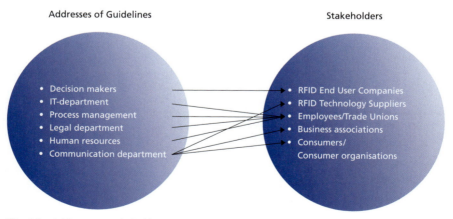

Fig. 4.8 Addressees, stakeholders and possible interactions

4.3 Quantitative Analysis of Guidelines

Table 4.7 Stakeholders and frequency of being considered

Stakeholders	Frequency of being considered
Market	9
Consumer	17
Employee/Trade Union	8
Media	2
NGOs	0

the interactions that might occur between the groups of addressees and the stakeholders. Decision makers, for example are potential addressees of guidelines and they might need information in these guidelines how to inform end user companies about RFID which can be cooperating companies in a closed loop system. Another example is the relationship between Human Resources and Employees. Human resources, as a guideline addressee, want to find information within a guideline of how to inform employees about the introduction of RFID technology.

As another result of the content analysis, stakeholders covered in the respective guidelines were summarised and are listed in Table 4.7. As this table shows, consumers are the group of stakeholders considered most often in the RFID guidelines analysed (17 out of 20 guidelines). Another relevant group of stakeholders is the market (nine out of 20 guidelines), followed by employees and trade unions. The media as a stakeholder seems to have a minor importance (two out of 20 guidelines) and is accordingly not notably considered. Non-Governmental Organisations (NGOs) are neither named nor described in any of the 20 analysed RFID guidelines.

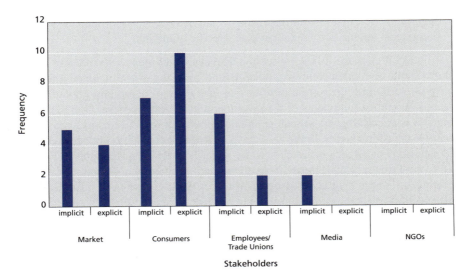

Fig. 4.9 Frequency of stakeholders being considered

The code rule for evaluating the consideration of stakeholders within the content analysis is to select whether the requirements are implicitly or explicitly fulfilled. The requirement is implicitly fulfilled, when the stakeholders are only named. An explicit fulfilment of the requirement means that the stakeholders are described in detail and there are, for instance, information strategies listed as well.

Figure 4.9 shows the frequency of considered stakeholders, differentiated into implicit and explicit fulfilment. The x-axis represents the stakeholders with whom the addressees may interact, further divided into an explicit or implicit fulfilment of the requirements. The y-axis represents the frequency of how often the stakeholders are considered in the analysed RFID guidelines. The diagram shows that stakeholders are considered more often implicitly, with the only exception of consumers. This is especially interesting as most of the RFID applications are still in the field of logistics and production i.e., in fields with no direct link to consumers. It would have been expected to find a higher consideration of the group of employees as they are involved or affected more often than consumers.

4.4 Establishing Guidelines Using the RFID Implementation Checklist

By using the RFID Implementation Checklist, companies, business organisations and all interested entities can create application specific guidelines for different addressees. The following three steps should be taken:

- Identify the specific application where RFID should be used and a guideline is needed, using for instance the RFID Reference Model as a tool to structure the different types of applications
- Identify the addressees with relevance for the specific RFID system
- Compile an application- and addressee-specific guideline

To make it clear the following section displays which steps a stakeholder would take to create an implementation guideline for a company or entity, helping them to integrate RFID into their organisation. The first stage of compiling a guideline is establishing the requirements of the system and its overall purpose. As a next step, it has to be clear who the guideline should be written for. Addressees can be for example decision makers, in the case that a strategic decision has not been reached. If a decision has already been made to set up such an RFID system, the IT Department, the Process Management Department and various other addressees require information, and it should be assessed which of these groups need information in that specific case and the guideline should be elaborated accordingly.

Decision Makers

The Decision Makers need to be informed as to how an RFID system can be efficiently implemented and how it can improve business performance. This group

4.4 Establishing Guidelines Using the RFID Implementation Checklist

does not usually require specific technical information, but rather an overview of the general impact of the system.

- **New Business Opportunities**: Here, potential business opportunities resulting from RFID implementation should be discussed.
- **Organisational Fit**: Potential impacts on business processes should be discussed roughly, and how RFID will affect already implemented processes.
- **Case Studies as a Benchmark**: It may be effective to present successful case studies to explain how the entire implementation project will function, and prove that RFID would make a positive impact on the organisation.
- **Data for Calculating a Business Case**: A clear outline of what information is needed to calculate a business case and what procedure should be followed or what kind of tools could be used should be included.
- **Cost/Benefit Analysis**: Detailed Cost/Benefit Analysis of similar use cases should be provided so that clear implications for the company can be derived and assistance on how to calculate own cost/benefit analysis.
- **Time Frame**: A Time Frame for the project should be established.
- **Resources Needed for Implementation**: Financial and human resources required need to be detailed.
- **Risk Assessment**: A specific description of potential risks needs to be elaborated for decision makers.
- **Measurability of Economic Success**: From existing use cases, benchmarks for the economic success can be established, taking into account current market conditions, company expectations, growth forecasts etc.
- **Non Monetary Positive Effects**: It may be important to highlight to decision makers what kind of additional benefits may be achievable by introducing the RFID system, such as being regarded as a very innovative company, boosting company image and reputation.
- **Legal Aspects**: Here, legal implications should be discussed roughly, such as employee rights, data protection and frequency restrictions.
- **Overview of Current RFID Technology**: An outline of the current events in the field of RFID technology could be useful to help the decision makers more aware of the technology and its current status.
- **IT Security**: Implications of the RFID system on IT security should be discussed.
- **Technology Requirements**: A brief overview of the technological requirements listing the proposed types of tags, readers and middleware could be discussed here.
- **Alternate Technologies**: A comparison of RFID technology to other Auto ID technologies, and a list of advantages and disadvantages.
- **Outlook, Future Developments, Roadmap**: An outlook on future developments will enable an effective timeline for implementation to be made.
- **Data Sharing Systems**: Issues concerning data sharing systems and the respective access rights should be discussed.

IT Department

The IT department needs to be informed of the technological requirements of the RFID system. The IT department will then be able to integrate the RFID system into the existing IT systems. IT departments need information about the following topics:

- **Standards**: Information is needed of which standards (e.g. data standards, air interface protocols) will be used.
- **Data Storage**: It has to be described what kind of data storage will be used, i.e. is all data stored on the tag or is only an ID stored on the tag and additional information in a database, or a combination.
- **Tag Technology**: The type of tag will need to be considered depending upon the application. As an example, this might be a tag which can withstand low temperatures or that is very robust may need to be chosen. It is important to list different types of tag which are suitable for the project and list considerations on the tag interoperability with other RFID systems.
- **IT Security**: IT security is an important issue with all IT systems, however if the new system handles sensitive data, then it may require additional security measures which have to be considered and properly addressed early on in the development phase.
- **Reliability**: Case studies and testing results of similar applications may be needed to evaluate the reliability of the system. Failsafe measures such as backup data storage and uninterruptible power supplies can be discussed here.
- **HW/SW Architecture**: An analysis of existing systems (hardware and software) will need to be conducted to establish how the RFID system can be integrated into existing systems.
- **Networking**: The RFID system will need to be integrated into the existing company network and IT processes.
- **Frequency Restrictions and International Interoperability**: Clear information about the frequencies used has to be given and frequency restrictions and international interoperability have to be discussed.
- **Interferences**: Existing radio transmission equipment must be tested to ensure that the RFID system will not interfere or be interfered with by existing equipment.
- **Interfaces**: It has to be described what kind of interfaces the RFID system will require.
- **Co-existence or Migration from classical Auto ID to RFID**: Assistance with deciding whether an existing classical Auto ID system should co-exist with the RFID system or how the migration should be conducted.
- **Deactivation**: In some cases it might be necessary to deactivate RFID tags. In such cases, deactivation possibilities need to be described.

4.4 Establishing Guidelines Using the RFID Implementation Checklist

Process Management

The process management department is in charge of optimising processes, and therefore it is concerned with how RFID can be implemented either alongside or in place of existing processes.

- **Process Descriptions of Other Use Cases**: Descriptions of processes of existing use cases to help process managers to understand how existing processes need to be adapted.
- **Attachment of Tags to Item**: Descriptions of how tags can be affixed to items correctly.
- **Awareness of Technical Properties of Different RFID Techniques**: Basic details about the technology and potential scenarios in the company or entity.
- **Co-existence or Migration from classical Auto ID to RFID**: Assistance with deciding whether an existing classical Auto ID system should co-exist with the RFID system or how the migration should be conducted and what are the respective effects on the existing processes.
- **Environmental Conditions**: Potential environmental conditions need to be considered which may affect the functionality of the tag. These could be issues such as temperature, pressure, or humid conditions.
- **Standards**: Standards regarding processes should be discussed here, including lists of potential standards which could be used.
- **Known Fields for Potential Improvement of Processes**: Use cases can be depicted to show how RFID can improve processes.
- **Outlook, Future Developments, Roadmap**: An outlook on future developments will enable an effective timeline for implementation to be made.

Legal Department

The legal department is concerned with the legal issues which may arise from using RFID. This may include internal legal issues such as rights for employees, or external factors, such as data protection of customer information.

- **Depicting Legislative Process Concerning Data Protection**: Information should be detailed here regarding the potential legal implications which may arise from implementing an RFID system.
- **Legislative and Self Regulative Compliance**: Information about relevant legislation and existing self regulation should be detailed.
- **Guidance When Written Consent for Data Collection is Necessary**: Guidance if written consent for data collection is necessary or on how to assess if written consent is necessary should be outlined here.
- **Explaining Legitimate Purpose for Data Collection**: Explanation of the reason for data collection to ensure that all data protection laws are complied with.

HR Department

The Human Resources department is concerned with the impacts that the new system will have on the employees of the company.

- **Overview of Current RFID Technology**: A brief outline of the current events in the field of RFID technology could be useful to raise awareness of the technology and its current status.
- **Scope and Purpose of Data Collection**: Outline what specific data will be collected, and the reason for its collection.
- **Health Issues**: Information regarding potential health impacts for employees.
- **Compliance with Employee Rights**: Employee rights must be addressed to ensure that no rights will be infringed upon by the new system, such as health and safety or data protection rights.
- **Influence on Employment**: Potential impact on organisational employment structure should be considered.
- **Training for Employees**: Training programmes for employees to work with the new technology should be described.

Communication Department

The Communication Department is concerned with all types of communications, both internally and externally. This could include information for employees regarding the new system, or press releases and dissemination of public information on RFID.

- **Information for Stakeholders**: Plans to inform stakeholders about the technology that should be deployed This includes providing stakeholders with information regarding issues such as the impacts the technology may have on privacy issues, environmental impacts and health issues.
- **Communicating Security Measures to Generate Trust**: The safety and security measures implemented should be explained to stakeholders to attempt generating trust.
- **Strategic Communication Campaign**: A strategic communication campaign can be outlined which can provide stakeholders with appropriate information regarding the RFID system.

Using this addressee-specific checklist, the writers of RFID guidelines can proceed and set up application-specific guidelines addressing the informational needs of different groups involved when setting up and using an RFID system.

4.5 Conclusions

4.5.1 Relevance of Existing RFID Guidelines

The term "RFID technology" does not refer to a single technology or single application solution. Based on the fundamental principle of transmitting data via radio waves, a vast number of different technologies, frequencies, and application scenarios can be compiled, which would all fit under the broad heading of RFID. Using RFID for the maintenance of railway cars, for instance, does not have a lot in common with using RFID for mobile ticketing solutions or to prevent counterfeiting of drugs.

Therefore, guidelines for implementing an RFID system have to be application-specific. Any attempt to draft guidelines aiming at covering several RFID applications or to give generalised guidance, will most likely prove to be insufficient in helping RFID users to implement RFID systems. General statements will be too broad and therefore of no relevance in real RFID use cases – even for small scale applications, technical requirements and system configurations may differ greatly, calling for an application-specific approach.

In order to assess the quality of existing guidelines, an in-depth content analysis has been conducted. It shows that only 20 out of 71 documents used as a basis for the analysis met the definition of "guideline". Thus 51 documents were only of limited practical use for companies that require advice regarding aspects that need to be considered when implementing an RFID system.

Regarding the level of fulfilment of guideline requirements, the median fulfilment was about 30%. The highest median fulfilment could be identified for the information requirements of communication departments. This shows that although RFID is still a developing technology with most of its possible applications still in research or pilot phases, it has gained considerable public attention and has sparked a lively and in some respect also controversial debate among stakeholders. This is one reason why communication plays a vital role in successful RFID implementations.

Following the assessment that every RFID application affects internal and external stakeholders, the guidelines have also been analysed in terms of how well they address concerns and requirements of relevant stakeholders. One of the major results was that the stakeholder group which is most commonly referred to is the end consumer – another indicator for the major public interest in RFID in recent years.

Generally speaking, the guidelines vary greatly in quality. Some of the documents analysed have shown to be sufficient to guide companies in actual RFID implementation efforts; however, most of them would have failed in doing so, because their scope was too broad, lacking vital detailed information needed for successful implementation.

Table 4.8 SWOT analysis of RFID implementation guidelines

Strengths	Weaknesses
Overview of possibilities	Too general or too theoretical to be useful
Sound basis for decision-making	Confusion about conflicting recommendations between different guidelines
New ideas, broader picture	
Best practice examples	Information often of no practical value
Opportunities	Threats
Saves time in planning	False sense of security, no incentive for self-analysis
Support to assess ROI (return on investment) in advance	Not applicable to the technology reality and its rapid development
Benchmarking and goal definition for implementation	Over complexity

Apart from the findings of the content analysis, the expert discussions and the relevant articles and publications have highlighted the general advantages and disadvantages of guidelines (see Table 4.8). Guidelines can serve as a sound basis for decision-making and provide best practices for a wide range of RFID applications. Based on a range of application-specific guidelines, the exchange of best practice examples for new RFID applications can be facilitated. Guidelines can help to save time and minimise business risks, however, it is important that the guidelines used are indeed application-specific. If this is not the case, a false sense of security on the part of decision makers and technical personnel might be the result if the information provided in a guideline does not reflect the specific circumstances and requirements for the respective RFID application. This stems from the fact that, as stated above, guidelines are often too general and quickly outdated due to the rapid technological progress in the field of RFID. On the other hand, some of them might also be too complex for first-time adopters.

4.5.2 The RFID Implementation Checklist – next Steps

As stated above, due to a broad scope and lack of sufficient information, existing guidelines are often of limited practical use for RFID implementation in user companies and other entities. More generally speaking, the entire concept of having commonly applicable guidelines for the technology as a whole has to be questioned.

However, there is no doubt that sound guidance is necessary in order to consider all challenges and aspects relevant in the course of RFID system implementation and use. Within the course of CE RFID, an RFID Implementation Checklist has been elaborated (see Chap. 4.1.2). This checklist attempts to list all relevant points that have to be considered by different addressees within the user com-

4.5 Conclusions

pany. This checklist therefore serves as the basis for the elaboration of application-specific guidelines for RFID projects within companies or other user entities (see Chap. 4.4).

One example of the activities already underway is the current discussion process initiated by the German BSI (Bundesamt für Sicherheit in der Informationstechnik, Federal Office for Information Security), which aims at publishing technical guidelines for different RFID applications (BSI 2008). They also follow a step-by-step checklist approach guiding companies or entities that are planning to use RFID through the questions and topics concerning security that need to be addressed and considered in RFID adoption. The BSI example shows again the direct connection and interdependency of guideline and checklist. Checklists as an outset for individual guideline development are therefore a feasible way forward.

The RFID Implementation Checklist as a suggestion in how to find the balance between sufficient guidance on the one hand and the necessary flexibility for specific use cases on the other, should not be considered to be final. For ensuring that the list can be a sufficient basis, further discussion, regular updates and more field tests are necessary.

Considering the results of the analysis of RFID guidelines and the necessary steps required in this field to move forward in technology implementation we give the following recommendations to especially support authors when drafting new guidelines:

Recommendation: Guidelines have to be application-specific

Strengthen the case for an application-specific approach by highlighting the fact that the multitude of RFID applications and the resulting challenges are too vast in order to allow one-size-fits-all solutions.

Recommendation: Guidelines have to be addressee-specific

Different addressees have different fields of interest when an RFID system is set up, for example the IT Department needs different information in comparison to Human Resources. When drafting guidelines it has to be clear who the guidelines are written for to integrate the informational needs of these target groups.

Recommendation: Involve relevant stakeholders in the process of updating the RFID Implementation Checklist

Those organisations taking care of the updating of the RFID Implementation Checklist have to assure to speak for a critical number of RFID users and to be well established in the ongoing dialogue on RFID. This way they are able to integrate the relevant stakeholders into the discussion.

> **Recommendation:** Foster the use and the continuous development of the RFID Implementation Checklist

As a solution to the demands for guidance in the RFID field, checklists can be a way forward for the transparent and safe implementation of RFID applications. Regular updates are necessary to ensure usefulness of the checklist considering rapid technological progress. The updating process has to be conducted by organisations that have the necessary skills and capacities to do so and provide the scope and market power needed to ensure a widespread adoption of the RFID Implementation Checklist. Therefore, organisations such as EPCglobal or AIM could play an important role in this effort.

Chapter 5
Regulatory Framework

Standards and information for potential users are not the only prerequisites for technological growth. It is also important that the legislative framework is reliable and supports the development of a new technology. In this chapter we analyse, from a legal point of view, issues which affect RFID. These issues relate notably to privacy and security, the impact on health and environment, Intellectual Property Rights (IPRs) and RFID governance. Recommendations for regulatory actions are drawn for each topic.

5.1 Privacy

Privacy is probably the topic which receives the most attention. Unlike barcodes or magnetic stripe cards, RFID technology does not require line of sight contact, allowing the data stored in the tag to be read without any notice or previous action from the data subject. This is the reason why a number of privacy concerns have been raised over the last few years with regard to this new technology which is predicted to reach widespread implementation in the upcoming future. In most cases RFID applications do not involve the storage of personal information related to an individual (like name, address, date of birth) on a tag (Strüker et al. 2008) but only a unique identification number (like a barcode) and, therefore, do not involve privacy issues. However, some RFID applications may directly or indirectly enable the identification of an identifiable person and bring about the risk of their disclosure to unwanted parties. A case-by-case approach should therefore be adopted.

In this chapter we will provide a summary of the most relevant privacy-related legislation at the European level, and a brief analysis of the privacy principles and rules that form the basis of the data protection legal framework. From this general level we will then go into a specific analysis of the privacy impact of RFID, compiling the relevant legal documents and developing an application-specific view (illustrated by real cases) on the privacy concerns raised by RFID. The aim is to analyse the existing legal framework alongside actual RFID applications to conclude whether or not new laws might be necessary, and which are the alternative options to address privacy concerns.

5.1.1 Legal Framework

In addition to the RFID-related EU privacy legislation (see Table 5.1 for a non-exhaustive catalogue), the "Communication on the follow-up of the Work Programme for better implementation of the Data Protection Directive" (COM(2007) 87 final) concludes that there is no need for amending it, in view of the current technological status. Furthermore, the "Communication: Promoting Data Protection by Privacy Enhancing Technologies (PETs)" (COM(2007) 228 final) sets up three main objectives in the way forward for the use of PETs: development, use availability and encouragement of consumers.

Regarding RFID in particular, in 2006 the European Commission carried out an online public consultation on the primary concerns of the citizens concerning the technology. The process ended in March 2007 with the release of the "Communication from the Commission regarding Radio Frequency Identification (RFID) in Europe: steps towards a policy framework" (COM(2007) 96 final, SEC(2007)312). According to the Commission, "further development and widespread RFID deployment could further strengthen the role of information and communication technologies (ICT) in driving innovation and promoting economic growth. Already today, Europe is a leading region in RFID-related research and development, not least thanks to the support of the European research programmes". Additionally, the Communication identifies a number of topics for which the question of adequacy of the legal framework may be raised. This chapter has taken in account the issues proposed by the Commission in the analysis of the relation between RFID and legislation. A Recommendation on privacy and security aspects raised by the RFID technology should be further adopted by the European Commission in the second-half of 2008.

The "Working Party on the protection of individuals with regard to the processing of personal data" (so called "Article 29 Working Party"), set up under the Data Protection Directive, also adopts documents giving guidelines related to data protection (e.g. "Opinion 4/2007 on the concept of personal data", 01248/07/EN – WP 136). On January 19th, 2005, the Article 29 Working Party published in particular a working document on data protection issues with regard to RFID technology (10107/05/EN WP 105). The working document is aimed at a) providing guidance to companies deploying RFID on the application of the basic principles set out in the directives, and b) providing guidance to manufacturers of the technology as well as RFID standardisation bodies on their responsibility towards designing privacy and compliant technology. It analyses RFID technology and its implications with regard to data protection matters, studying applications, privacy and security issues, and technical solutions.

5.1 Privacy

Table 5.1 RFID-related EU privacy legislation

Name	Summary
Consolidated versions of the Treaty on European Union and of the Treaty establishing the European Community (OJ C 306 17.12.2007)	Art. 6 of the Treaty on European Union guarantees the respect of the fundamental rights laid down by the European Convention for the Protection of Human Rights and Fundamental Freedoms of 1950 (Art. 8), and Art. 286 deals specifically with data protection.
Treaty of Lisbon amending the Treaty on European Union and the Treaty establishing the European Community (OJ C 306 17.12.2007)	Not yet in force. It will introduce specific provisions concerning human rights and fundamental freedoms (Arts. 1.8, 2.29).
92/242/EEC: Council Decision of 31 March 1992 in the field of security of information systems (OJ L123, 8.5.1992, p.19)	Sets up the basic framework for developing a protected environment in the field of data storage and processing. It represents the starting point for the following legal texts in the field of data protection and data security.
Directive 95/46/EC of the European Parliament and of the Council of 24 October 1995 on the protection of individuals with regard to the processing of personal data and of the free movement of such data (so called "Data Protection Directive") (OJ L 281, 23.11.1995, p.31)	The directive establishes the main principles for lawful processing of personal data; it is the cornerstone of the European Data Protection legal framework.
Regulation (EC) No 45/2001 of the European Parliament and of the Council of 18 December 2000 on the protection of individuals with regard to the processing of personal data by the Community institutions and bodies and of the free movement of such data (OJ L 8, 12.1.2001, p.1)	The regulation provides rules to process personal data within the different European institutions and bodies, and establishes an independent body for supervision of their application.
Directive 2002/21/EC of the European Parliament and the Council of 7 March 2002 on a common regulatory framework for electronic communications networks and services (Framework Directive)	The directive establishes a harmonised framework for the regulation of electronic communications networks and services.
Directive 2002/58/EC of the European Parliament and of the Council of 12 July 2002 concerning the processing of personal data and the protection of privacy in the electronic communications sector (so called "ePrivacy Directive") OJ L 201, 31.7.2002, p.37 (amended by Directive 2006/24/EC)	The directive deals with a number of issues such as data retention, the use of cookies and the inclusion of personal data in public directories. Its scope is limited to "the processing of personal data in connection with the provision of publicly available electronic communications services in public communications networks."

In order to complete the picture of applicable regulations, industry guidelines and self-regulation must also be discussed. Guidelines, in this project, are seen as implementation guidelines, i.e. "domain-specific documents assisting companies and organisations when implementing and using RFID systems with regard to possibly affected stakeholders. Such guidelines need to address specific concerns of entities within companies and organisations that implement or use RFID systems" (see Chapter 4.2.3). Self-regulation, on the other hand, is understood within the project as regulations which companies or entities impose on themselves and abide to. According to this meaning, self-regulation would be placed between guidelines, which are simply advice, and law, which must be adhered to. In any case, self-regulation and legislation are two sides of the same coin; both are intended to ensure that privacy is respected when implementing RFID. To be efficient, self-regulations need to be coherently enforced by the industry: while admitting that by their own nature they cannot be legally enforced, self-regulations would make little sense if they are not respected by their addressees. For a deeper analysis of guidelines and RFID see Gampl et al. (2008a).

5.1.2 Data Protection Principles and the Definition of Personal Data

As we have already stated, Directive 95/46 ("Data Protection Directive" hereinafter) laid down the general basis of the European data protection legal framework. Its articles (further developed or modified by other legal texts) introduced the principles and common rules that should be followed to ensure lawful processing of personal data. Furthermore, the deefinition of personal data contained in Art. 2 of the said directive is the key to determine which applications fall within the scope of the text. In the following paragraphs the content of the most prominent articles within the Data Protection Directive will be briefly described, in order to establish the basis for the subsequent study of RFID and privacy.

5.1.2.1 General Content

The scope of the Data Protection Directive, stipulated in Art. 3, is restrained to the processing of personal data carried out entirely or partially by automatic means, and to the non-automatic processing of personal data "which form part" or "are intended to form part" of a filing system. This means that an application dealing with data not qualified as personal falls out of the scope of the Data Protection Directive. The text is applicable to all controllers established in Member States or territories where Member State's law is applicable (Art. 4). According to the Data Protection Directive, a controller is "the natural or legal person, public authority, agency or any other body which alone or jointly with others determines the purposes and means of the processing of personal data (…)" (Art. 2. (d)).

5.1.2.2 Data protection Principles

The Data Protection Directive lays down a number of principles that Member States shall determine more precisely, how the processing of personal data should be carried out in order for it to be lawful. These constitute the data protection principles and are covered by Arts. 5 to 8. The conditions, under which a processing of personal data qualifies as lawful and legitimate, break down into three different types: data quality, legitimate processing and special categories of processing.

In order to ensure the quality of the personal data (Art. 6), the controller should ensure that personal data is:

- Processed fairly and lawfully (the processing shall comply with every legal provisions concerning data protection);
- Collected for specified, explicit and legitimate purposes: when the processing proves to be incompatible with those purposes, it shall not be allowed. Exemptions can be found in historical, statistical and scientific purposes;
- Adequate, relevant and not excessive in relation to those purposes;
- Accurate and kept up to date;
- Kept in a form which permits identification of data subjects for no longer than necessary;

The processing of personal data will be considered legitimate only if one or more of the following criteria are met (Art. 7):

- The data subject has unambiguously given his or her consent. Forced, unclear or non implied consent shall not qualify;
- Processing is necessary for the performance of a contract to which the data subject is party, or at the pre-contractual level;
- Processing is necessary for compliance with a legal obligation of the controller;
- Processing is necessary in order to protect the vital interests of the data subject;
- Processing is necessary for the performance of a task carried out in the public interest or in the exercise of official authority, conferred to the controller or to an authorised third party;
- Processing is necessary for the purposes of the legitimate interests pursued by the controller or by the third party or parties to whom the data is disclosed (invalid if the data subject's rights and freedoms could be harmed by such processing).

According to Art. 8 concerning special categories of processing, the processing of personal data shall be prohibited (with exceptions such as medical purposes, for example) if it reveals or concerns racial or ethnic origin, political opinions, religious or philosophical beliefs, trade-union membership, health or sex life; unless:

- the data subject has given his or her consent,
- processing is necessary to protect the vital interests of the data subject or
- data has been made public by the data subject.

5.1.2.3 Other Rules under the Data Protection Directive

- The data subject should be given certain information, such as the identity of the data controller, the purposes of the processing or the existence of certain rights (access, deletion, etc.).
- The data subject has the right to access, rectify and to block access to his or her data. They are also entitled to object at any time to the processing of data relating to them.
- No one can be legally affected by a decision taken on the basis of a pure automated processing of data.
- The controller shall make sure that the processing is secure and confidential. Furthermore, they "must implement appropriate technical and organisational measures to protect personal data" (Art. 17).

For a deeper analysis of the Data Protection Directive please refer to the extended project report on European RFID legislation (Kruse et al. 2008).

5.1.2.4 The Definition of Personal Data

Article 2 (a) of the Data Protection Directive defines personal data as follows:

> Personal data shall mean any information relating to an identified or identifiable natural person ('data subject'); an identifiable person is one who can be identified, directly or indirectly, in particular by reference to an identification number or to one or more factors specific to his physical, physiological, mental, economic, cultural or social identity.

The concept of personal data is the key provision to determine whether a particular technology application falls within the scope of the Data Protection Directive. Only those applications which involve personal data shall comply with the above principles regarding processing.

However, *what* should be understood as personal data has been the subject of an intensive debate; in this section we will try to clarify the concept so as to establish a basis to analyse the different kinds of RFID applications, and whether they involve personal data or not.

According to the Article 29 Working Party the analysis of the definition should be focused on the first sentence of the paragraph of Art. 2(a): "any information relating to an identified or identifiable natural person" (Art. 29 Working Party, WP136, 2007).

"Any information" shall include subjective and objective information, not necessarily true or proven, made available in any form (e.g. e-mail), and including biometric data ("biological properties, physiological characteristics, living traits or repeatable actions where those features and/or actions are both unique to that individual and measurable, even if the patterns used in practice to technically measure them involve a certain degree of probability"). The most important feature about this kind of data is that it could serve as a link between a particular individual and certain information.

5.1 Privacy

For information to be classified as "relating to" someone, the information shall be about a person (e.g. name, address), or, in the cases that it is not about a person, it shall be possible to use it to take some actions over an individual or in a way that has some kind of impact on them. That information can be direct (when accessing X's hospital records, that information is "directly" related to X), or indirect (one gets to discover X's personal information by knowing X's car registration number and by that gets access to the car registration record including X's personal information).

The person shall be understood as "identified" when they are "distinguished" within a group of persons by using (a) certain characteristics to identify them, be it name, date of birth, or eye colour. The subject will be classed as "identifiable" when it is simply "possible" to identify them within a group, hence the suffix "-able" (Art. 29 Working Party, WP136, 2007).

The individual can then be identified directly (e.g. by name) or indirectly (e.g. by passport number, car registration, or a combination of records which allows an individual to be identified). As to determine whether a person is identifiable or not, Recital 26 of the Data Protection Directive states that:

> "whereas to determine whether a person is identifiable, account should be taken of all the means likely reasonably to be used either by the controller, or by any other person to identify the said person."

In accordance with Recital 26 of the Data Protection Directive, the Art. 29 WP considers that *"a mere hypothetical possibility to single out the individual is not enough to consider that person as "identifiable""* (Art. 29 Working Party, WP136, 2007). If, taking into account *"all the means likely reasonably to be used by the controller or any other person"*, the possibility to identify an individual through the data involved with a particular technology application does not exist or is negligible, the person should not be considered as "identifiable", and the data concerned should not be considered as "personal data" (Art. 29 Working Party, WP136, 2007).

The Art. 29 WP adopted a pragmatic approach regarding the assessment of whether a person is "identifiable" or not by stressing that *"the criterion of 'all the means likely reasonably to be used either by the controller or by any other person' should in particular take into account all the factors at stake. The cost of conducting identification is one factor, but not the only one. The intended purpose, the way the processing is structured, the advantage expected by the controller, the interests at stake for the individuals, as weell as the risk of organisational dysfunctions (e.g. breaches of confidentiality duties) and technical failures should all be taken into account. On the other hand, this test is a dynamic one and should consider the state of the art in technology at the time of the processing and the possibilities for development during the period for which the data will be processed. Identification may not be possible today with all the means likely reasonably to be used today."* (Art. 29 Working Party, WP136, 2007).

In the case where the information was collected with the purpose of identifying individuals, the person shall be considered as "identifiable".

However, where identification of the data subject is not the purpose of the processing, the technical measures to prevent identification have to play a very important role. As explained by the Art. 29 WP, *"putting in place the appropriate state-of-the-art technical and organizational measures to protect the data against identification may make the difference to consider that the persons are not identifiable, taking account 'of all the means likely reasonably to be used by the controller or by any other person' to identify the individuals. In this case, the implementation of those measures is not the consequence of a legal obligation arising from Art. 17 of the Directive (which only applies if the information is personal data in the first place), but rather a condition for the information precisely not to be considered to be personal data and its processing not to be subject to the Directive."* (Art. 29 Working Party, WP136, 2007).

As for the term "natural person", it should be assumed that data related to dead people should not generally be considered to fall within the scope of the Data Protection Directive. Legal persons are also, in principle, excluded although there are some exemptions.

5.1.3 RFID and Data Protection Legislation: a Case Specific Approach

Having outlined the general legal provisions governing data protection, the question is now to apply those rules to RFID technology applications on a case-by-case basis.

As any technology involving data, RFID technology falls, in principle, into the scope of the Data Protection Directive. However, the provisions of the Data Protection Directive will only apply to a specific RFID application when the RFID application concerned can be considered as involving "personal data". In this respect, we have classified below the different types of RFID applications, depending on whether or not they involve personal data as provided by Art. 2(a) of the Data Protection Directive.

5.1.3.1 RFID Tags Containing no direct, indirect or potential Identifiers

Most RFID tags only contain information (usually just a unique identification number) related to the product concerned, purely used for organisational and production effectiveness purposes within a company or throughout the supply chain. In this case, the individuals in contact with those tags are only the employees of the companies concerned and, if the application concerned would allow identification of a particular employee, the principles governing processing of personal data apply and any action challenging the employee's privacy is prohibited (see below Chap. 5.1.3.5).

For RFID logistics and manufacturing applications, information can not therefore in principle be directly related to a specific data subject or lead indirectly to the identification of a specific data subject. For example, in the case of RFID devices used for preventing the use of counterfeited or damaged components in aeroplanes and vehicles, RFID is used for ensuring safety and security in high risk environments, and therefore identifiers are given to the components of either vehicles or aeroplanes. A tag can be placed on each part of the craft, and the tag stores a unique identification number that, when interpreted by a reader, allows the employees of the company concerned to know where the part should be placed, and if it is adequate and authentic. In this process, no data subject outside the company concerned is involved and no personal data is stored on the tag, therefore, there is no indirect or direct personal data involved. Another example can be found in the area of closed loop logistics, with companies tagging their packaging containers for control purposes. Again, no personal data is involved, since the containers are classified with numbers and only for management reasons.

As a result, it is generally agreed that the Data Protection Directive's provisions regarding the process of personal data are not applicable to logistics and manufacturing RFID applications as no personal data is involved.

Once the consumer comes into contact with the tagged product, some privacy concerns may, however, be raised due to the possible link between the unique identification number stored on the tag and some personal data of the consumer stored in a database (see Chap. 5.1.3.2). This is in particular the case for applications in the retail industry even though the tags do not store personal data of a data subject, like tags used for logistics or product maintenance and quality control purposes do.

5.1.3.2 RFID Tags which store Information that could be linked to Personal Data

This is the situation where "*indirect*" identification might take place: the tag contains a unique identification number that can be reasonably linked to a particular person and/or personal data.

In the retail sector for instance, tags can be currently placed on products in order to make the work of the staff easier and to improve the logistics of the store and customer's service. They do not have the purpose of identifying the retailer's customers.

No personal data of an individual is stored on the tags and read alone, they contain a unique identification number enabling the recognition of the product and possibly some further information regarding the product concerned (e.g. expiration date, country of origin).

If a product is tagged and read in the store (regardless if the tag is deactivated before leaving the store or not) and if the customer is not identified by means of a loyalty or credit card for instance in the store (usually at the check out point), the customer cannot then reasonably be identified through the RFID tag by the retailer or after having left the store, due to the fact that their personal data has not been disclosed and could not be potentially linked to the unique identification number on the tag.

However, when linked to the personal data of an individual extracted from a database, related for instance to the customer's loyalty card or credit card, the unique identification number stored on the tag can, in principle, be potentially used to identify an individual. Nevertheless, in practice, identification through a credit card's database is complicated since the process is encrypted, and the data stored is secured according to applicable data protections laws to be used only for specific purposes to which a user expressly consented.

Therefore, only linking between a unique identification number on a tag with the personal information of an individual stored in a database makes a person "identifiable" (as explained above), which means that, in such a case *"taking into account all the means likely reasonably to be used by the controller or by any other person to identify the individuals"*, the RFID application concerned may be considered as involving personal data, and, thus, be subject to data protection rules. Furthermore, according to the Art. 29 WP, the RFID applications potentially enabling the identification of persons, but of which the purpose is different (such as logistics enhancement), particularly require *"appropriate state-of-the art technical and organisational measures to protect the data against identification"* (see above WP136, 2007).

In contrast, RFID application may in other sectors, like in healthcare, have the purpose of identifying persons, which makes the application of data protection rules more obvious. In the trial held in the AMC Hospital in Amsterdam for instance, patients received RFID tagged bracelets on their arrival. The tags contain a unique identification number which is linked to the hospital's database where medical records are stored. Tags were also used to identify blood transfusion instruments and respective patients in order to avoid mismatches. Therefore, those types of applications should be considered as involving information relating to identifiable individuals and the processing should be subject to the data protection rules.

According to Recital 26 of the Data protection Directive and following the recommendation of the Art. 29 WP (WP 136, 2007), due to the large number of RFID applications, the analysis of whether or not a person can reasonably be considered as identifiable by a specific RFID application should, therefore, be done on a case-by-case basis, taking into account factors such as the cost of the identification, the purpose, the expected advantage for the controller, the interests of all the parties, the security device applied etc.

In its "Working document on data protection issues related to RFID technology" (WP 105, 2005), the Art. 29 WP mentions a number of hypothetical examples with regard to the profiling of consumers and inducing malicious and unlawful processing of personal data.

However, a number of factors should reasonably prevent those kinds of situations from happening. First of all, the current level of technology does not allow such tracking. Secondly, from a practical point of view, it is hardly feasible that a company could trace a customer by obtaining only one identifier.

5.1 Privacy

In addition, and assuming that the technology would be available, the profiling of individuals is already forbidden by law, and thus, any company using such techniques to track unknowing individuals would be committing a punishable criminal offence.

Furthermore, as with any other technology, RFID applications may be the object of malicious and unlawful usage from third-parties. However, an RFID application user having carried out a privacy impact assessment prior to the development of the RFID application and having installed state-of-the-art security, technical and organisational measures related to the privacy risk concerned, this user should be considered as having taken all reasonable measures to prevent all reasonable privacy risks linked to its RFID application. This user cannot be held liable for any malicious and unlawful usage of its RFID application by a third party and should not be prevented from implementing his or her RFID applications.

According to Art. 29 WP, guidelines on the compliance of the data protection requirements would help implementing RFID applications for the benefit of the industry and of the society alike.

5.1.3.3 RFID Tags which store Personal Data

In December 2004 the European Council adopted Regulation (EC) 2252/2004 introducing the so-called "European Biometric Passport" by 2005. However, due to significant delays, only some of the Member States had issued the passports containing a facial image on the tag on time, i.e. by 28 August 2006. By 28 June 2008, the Member States shall have two fingerprints added to the information contained in the chip. As already mentioned (Chap. 5.1.2.4), biometric data are considered as an additional category of personal data and are, therefore, covered by the Data Protection Directive and related legislation. Biometric passports are probably the most obvious example of an RFID application containing personal data.

RFID applications including tags storing personal data have to comply with the provisions of the Data Protection Directive and related legal instruments and are therefore covered by the existing legislation.

There are, however, limited examples where RFID applications with storage of personal data on the tag. In most of the RFID applications of today and tomorrow, the way to identify a person is by combining the unique identification number on the tag held by a person with a back-end database where personal data of the concerned individual is stored.

5.1.3.4 Applications of the Data Protection Principles to RFID

RFID technology, as any technology, principally falls into the scope of the Data Protection Directive. As explained above, however, the provisions of the Data Protection Directive will be applicable to a particular RFID application when it involves the processing of personal data according to Art. 3.1. of the Data Protection Directive. Consequently, the application of the provisions of the Data Protection

Directive should then be determined on a case-by-basis depending on whether the RFID application concerned involved personal data, making a person identifiable.

In particular, all data protection principles (see Chap. 5.1.2) imposed by the Data Protection Directive should be adhered to by the controller. In the case of RFID applications, the "controller" would be the user of the tag, who determines the purpose of that tag used in combination with the processing of the tag information to the reader and from the reader to other means, such as databases. The user is bound to the requirements of purpose limitation, proportionality and conservation principles laid down by Art. 6 of the Data Protection Directive. This means that both the controller and the manufacturer of the RFID tags shall structure the system in such a way that only the necessary data is collected and processed for specific purposes, and that their content is proportional to the purposes for which they were collected.

Concerning the legal grounds for processing, as provided by Art. 7 of the Data Protection Directive, the key element is the consent of the data subject. The processing of personal data is lawful and allowed provided that the data subject has unambiguously given his or her consent, except in some cases. Consent should be given freely, specifically and unambiguously. However, the Data protection Directive does not provide a specific method on how to grant that consent.

Besides, Art. 8.2 of the Data Protection Directive requires consent for the processing of sensitive data "explicitly", which may suggest that not in all cases "explicit" consent of the data subject is required under the Data Protection Directive. "Explicit" and "tacit" consent could then be differentiated depending on the type of personal data concerned. Furthermore, several authors have supported the idea that a "tacit" consent shall suffice in most of the cases and therefore, that opt-out shall be the general rule (Téllez 2002, Aparicio 2000, among others). The Art. 29 WP seems also to accept that opt-out provides "practicability and flexibility" (WP 131, 2007).

For RFID applications storing personal data on the tag or having the identification of persons as their purpose, the data subjects are, in practice, generally asked to give their consent explicitly.

For RFID applications not used for the identification of persons, but where a potential may exist that information is linked to an identifiable person, supplementary consent is usually put in relation with the deactivation of the tag, in addition to the necessary consent from the data subject for processing personal data. Two scenarios with regard to the granting of the data subject's consent could then been identified: "opt-in" (standard deactivation) and "opt-out" (deactivation on request). In the first case, the idea is that the data subject *should actively* give their consent specifically to the RFID application; this would imply, for example, that retailers using tagged items shall provide devices to deactivate all tags unless the customer gives his or her explicit consent for the tag(s) to remain active after leaving the premises. In the latter case, the tags would remain active unless the data subject expresses the desire of deactivating it. Today, the retailers having developed some RFID applications generally adopt deactivation on request by making deactivation device available to the customers before leaving the store.

On one hand, the risk and privacy impact assessment conducted by retailers has shown that adoption of an "opt-out" solution (deactivation on request) is the best option since it addresses the potential privacy risks reasonably raised by their RFID applications while ensuring their development and taking into account the current technologies available. Deactivation on request is, however, developed in combination with a good information system towards the consumer in order for them to make their cost/benefit assessment regarding tag deactivation, and then to make a choice in a fully informed way. Retailers also adopt all technical and organisation measures to ensure data security and do not illegally link RFID data with personal data.

On the other hand, the risk and privacy impact assessment performed by the retailers shows that the "opt-in" scenario (deactivation by default) would impose high technical and costly burdens on retailers and prevent the development of beneficial after-sales use cases for maintenance or food safety purposes for instance. In the case of a small or medium private retailer deploying no specific RFID application for its own use but selling tagged products (the manufacturers having tagged their products purely for logistical purposes), the generalisation of the "opt-in" scenario would oblige the private retailer to implement a RFID deactivation solution, even though it does not use RFID for its own use. Furthermore, considering that it is not possible for manufacturers to differentiate at the production level products to be sold to private retailers and to retail chains, and the state-of-the-art deactivation solutions which do not allow deactivation of all types of tags with one device, the practical and technological burdens linked to the deactivation by default seem not proportionate to the privacy risks raised by the RFID applications.

In view of the future expanded use of RFID in the retail sector, the upcoming Recommendation of the European Commission on privacy and security should especially give further guidance for the development of RFID applications by retailers, taking particularly into account the privacy of the consumers, the state-of-the-art technology, the practical circumstances of a retail environment, and the consumer benefits of after-sale RFID applications for maintenance or food safety purposes.

For all RFID applications storing personal data on the tag or enabling on purpose or not the identification of a person, two prerequisite conditions must be complied with: full information to the data subject and security of processing (i.e. taking all reasonable technical and organization measures preventing the identification of the data subject).

According to Arts. 10 and 11 of the Data Protection Directive, the data subject should be informed of a number of points, such as the identity of the data controller, the presence of RFID applications and the possibility that information could be read without any action from the subject. Information is the key when it comes to RFID and privacy. Several solutions have been proposed, from using pictograms to the handing over of notices for consumers (OECD 2008a). In cases where the applications make it impossible to provide complete information to the data subjects, signs such as those used for the CCTV (Closed Circuit Television), such as "RFID used here" have been considered.

Nevertheless, in view of the lack of general knowledge of the benefits and risks of RFID technology and applications, an extensive public information campaign on the purposes, effects, advantages and disadvantages for both the industry and the society of RFID technology could, in principle, play a major role in the process of familiarising the public with it, and could also favour its widespread implementation in full respect of privacy rights.

In addition, implementation of technical and organisational measures would make the RFID application in compliance with the requirements on the security of processing provided by Art. 17 of the Data Protection Directive and would enable the data subject to exercise his/her rights of access and to object as provided in Arts. 12 and 14 of the Data Protection Directive.

In general, security of processing should be addressed by means of Privacy Impact Assessment prior to the implementation of RFID applications (OECD 2008a). There are already today some technical guidelines for implementation and utilisation of RFID-based systems, such as those currently developed by the German federal office for information security (BSI), which explain the method to follow.

Today, the retailers who have developed some RFID applications generally adopt deactivation on request by making a deactivation device available to the customers before leaving the store.

According to the Art. 29 WP, where identification of the data subject is not the purpose of the RFID application but may potentially be possible, implementing technical measures to prevent identification plays a very important role since it should avoid the information to be qualified as personal data and its processing to be subject to the Data Protection Directive (Art. 29 Working Party, WP136, 2007).

Technical measures may offer a number of options to secure RFID devices in order to protect privacy: tags could be designed in order to limit the potential privacy risks of some RFID applications. This is what has been labelled as "Privacy by Design". Depending on the privacy risks linked to the specific applications foreseen, different designs of RFID devices would be available in order to provide the most adequate technical security. Linked with the notion of "Privacy by design" is the concept of Privacy Enhancing Technologies as provided by the European Commission ("Communication: Promoting Data Protection by Privacy Enhancing Technologies" COM(2007) 228 final). According to the PISA project (Privacy Incorporated Software Agent), PETs are *"ICT measures that protect privacy by eliminating or reducing personal data or by preventing unnecessary and/or undesired processing of personal data, all without losing the functionality of the information system"*. Some of the technical measures to implement data protection provisions are: techniques enabling visual indications of activation, data tracks to access personal data, "kill" and "sleep" commands, privacy bits (i.e. a bit placed on the memory of a RFID tag that determines whether the tag can be read by all readers or only authorised readers), clipping antennas, blocker tags or "Faraday cages" to ensure the right to erase or block the data (Kruse et al. 2008).

Information and security requirements are at the core of any RFID applications and should be treated carefully by RFID deployers in order to address privacy concerns linked to its RFID application in an adequate and proportionate manner. Implementation of any RFID application should be made only after an adequate Risk and Privacy Impact Assessment according to the applicable guidelines, if any.

5.1.3.5 RFID in Workplaces

Deploying tags in the workplace may increase privacy concerns for a number of reasons. One of them is the fact that the use of RFID is based on a non-balanced relationship, making it almost compulsory for employees to use RFID. Furthermore, even though an employee may not usually be located throughout the entire workplace, there is the possibility to trace them when passing readers (however only if specific personal data of the employee has been collected). The deployment of RFID in the workplace also means that certain indicators can be checked, i.e. levels of performance, time spent in the office or length of breaks.

Apart from the tracing of employees, the rest of the concerns already exist with current technologies (such as CCTV). In order to address the privacy concerns related to the use of RFID in the workplace, a number of options can be listed, such as the implication of Workers Councils in the implementation of technology (compulsory in Germany, for example), or the enactment of codes of conduct.

In this respect, the International Labour Organization is playing a major role in the introduction of RFID technology at work by carrying out extensive researches on the social, privacy and labour implications of the deployment of RFID. Although employees are covered (as any other person) by the extensive legal framework dealing with data protection, there are some scenarios where they might be exposed to further risks concerning their privacy more than the average citizen. In particular, the majority of cases show that the provision of personal data is somehow compulsory for the employees or even for a person to get to a job. It is considered as normal that, when working for someone, one has to give up some privacy (OPC 2008).

That is why guidelines such as "Protection of Workers' Personal Data – An ILO code of practice" or the recommendation adopted by the Privacy Commissioner of Canada ("Radio Frequency Identification in the Workplace: Recommendation for Good Practices" of 2008) are so important in these cases. The latter document advocates "taking a proactive stance in the development and deployment of new technologies" so as to "enhance privacy by ensuring careful and appropriate design and deployment of the technologies in a manner that anticipates and respects privacy concerns". The Privacy Commissioner of Canada is particularly concerned with the "secondary" uses that the deployment of RFID at workplaces might have, i.e. surreptitious surveillance of workers or tracking for purposes other than those accepted as "legitimate". In order to assess the issues that might arise in the context of RFID-related privacy issues for employees, the document recommends conducting Privacy Impact Assessments (taking into account the concept of personal information and the reasonableness of personal data collection among others). The document establishes a number of good practices for employers when implementing RFID at work, such as having an accountable person within the organisation, identifying the purposes for implementation, guaranteeing the fair and informed consent of employees, limiting collection of personal data, as well as the use, the disclosure, the retention, and the updating of personal data.

As a corollary, it is important to point out that when implementing RFID technology at the workplace, employers should ensure that all labour regulations are being complied with, and proportionality applied at every level.

5.1.4 Conclusions

RFID is a technology that could place Europe in a very advantageous position if developed in the right way. It has many applications, ranging from healthcare to logistics, from public transport to libraries, which would make people's lives easier and improve efficiency in a number of industrial and service processes. Nevertheless, this potential could be harmed if potential privacy concerns raised by some RFID applications are not tackled with adequate tools. According to the OECD, "RFID technology is at a stage of development where privacy and security have been identified as challenges for its widespread adoption (...) RFID security and privacy should be an urgent priority for all stakeholders in order to prevent large scale opposition by consumers and individuals, and facilitate the successful roll-out of future RFID systems." (OECD 2008a)

RFID technology is, however, already covered, as any other technology, by the existing data protection legal framework. Given the current technological status and the current development of RFID applications, the application of the existing data protection rules adequately addresses privacy concerns which may be raised by some RFID applications.

Considering the large number of different RFID applications, the OECD recommends that firms who use RFID perform (before the implementation of the technology) a Privacy Impact Assessment in order to determine whether the concerned applications involve personal data or not (OECD, 2008a). The Privacy Impact Assessment should be based on the principle of reasonableness, i.e. it should assess the reasonable privacy risk linked with a RFID application provided that all adequate and reasonable technical security measures have been implemented.

On these grounds, the following conclusions can be stated:

> **Recommendation:** The existing data protection legal framework is adequate for RFID technology applications

5.1 Privacy

As any other technology, RFID shall comply with the Data Protection Directive and related legislation and with Member States' national data protection laws.

The privacy impact of RFID technology should be evaluated application by application. If RFID applications involve personal data, they are covered by the existing data protection legislation, guided by the principles of a) technology neutrality and b) informed consent of the individuals (OECD 2006). The current legal framework is considered as flexible enough to cope with further developments of RFID applications (Holznagel et al. 2006).

Moreover, a specific data protection and/or privacy regulation applicable only to RFID technology would hamper the development of RFID applications by the industry and especially by SMEs. This is particularly true, as item-level applications in open supply chains, that are at the core of the current privacy debate are, not in widespread use today, and would be in place only in a mid-term perspective (Strüker et al. 2008).

> **Recommendation:** Enforcement of the current data protection legal framework is fundamental

Considering the legal requirements to comply with in order to address the privacy concerns raised by some RFID applications, the condition for securing the deployment of RFID applications in full respect of the right of privacy is to make sure that the existing data protection legislation is applied.

Continuous dialogue between the European Commission and the Member States should be encouraged in order to monitor the application of data protection legislation related to RFID applications and to avoid radical differences in the implementation among the 27 Member States. Such differences would hold back the development of EU-wide RFID applications.

In this respect, in principle, the future recommendation addressed to Member States regarding privacy and security related to RFID should adequately help to have a consistent enforcement and interpretation of the data protection principles within the EU.

> **Recommendation:** Public information on RFID is crucial

In accordance with the rules governing data protection, when personal data is involved, complete information should be given to the individual in order to enable them to agree or not to have their personal data processed. The consent of the data subject constitutes the legal basis for processing personal data in the majority of cases. This consent should be unambiguous, freely given, specific and informed. Therefore, informing the consumer in a clear and understandable manner of the consequences of having an active RFID tag is an additional measure that can address most of the privacy concerns. In any case, it should be taken into account that the need for information goes along with the question of whether the tag used in the given environment contains personal data or not.

If RFID is to be implemented on a broad scale, it needs to have the consumers' support, and that can only be achieved by informing them in a proper manner about the advantages and disadvantages of the technology, which is currently largely unknown by the public. Only complete information will allow the individual to understand the benefits of RFID technology and of its application in their day-to-day life, while also answering questions about the potential legitimate risks linked to their privacy.

In this respect, a European and/or nationally funded public campaign will be an efficient tool to familiarise the general public with the technology. Examples on how media coverage could be carried out can be found on TV or on the internet, with "EU Tube" hosting a complete video on the subject and web pages such as www.discoverrfid.org or www.rfidabc.de which aim at explaining how the technology works in an understandable and entertaining way. A differentiation between the diverse RFID applications and case-by-case information about their potential impact on privacy shall be provided to the general public.

> **Recommendation:** The protection of personal data should be ensured without imposing unreasonable burdens upon the RFID users

According to the legal data protection rules, in addition to complete information to be given to the individual and to the implementation of all reasonable security measures, the data subjects must be enabled to give their consent for the processing of their personal data.

Taking into account of all the reasonable means likely to be used by the RFID user or by any other person to identify the individuals, the method applied by the RFID user to enable the data subject to agree or not with the processing of their personal data should be adequate and proportionate to the privacy risks involved but should enable freely given, specific and unambiguous consent.

The RFID user would be able to decide which method to apply for enabling the consent of the data subject by doing privacy and risk assessments prior the development of its RFID application.

If consistent public information is provided and all reasonable technical and organisational measures are applied to secure data, the "opt-out" method should be recognised as complying with all legal requirements, without hindering the development of the RFID technology.

In the field of RFID, consent from the data subject to specific RFID applications is often put in relation with deactivation of the tag. Two scenarios with regard to the granting of the data subject's consent can then been identified: "opt-in" (standard deactivation) and "opt-out" (deactivation on request).

In the retail sector for instance, privacy and risk impact assessments show that an "opt-out" policy would, on one hand, address the privacy concerns proportionally to the current risks foreseen and, on the other hand, allow the development of RFID applications, take into account of the technical and structural problems related to the deactivation of the tags by default and allow after-sale services enabled by the tags related, for instance, to recycling and anti-counterfeiting purposes.

In this respect, the future Recommendation of the European Commission regarding privacy and security related to RFID should give guidance to RFID users for applying the most adequate method for obtaining, when required, the consent of the data subject.

> **Recommendation:** Technical measures should be used to ensure personal data security

Implementation of any RFID application should be done only after an adequate Risk and Privacy Impact Assessment according to applicable guidelines if any.

Technical and organisational measures can offer a wide range of solutions to address privacy concerns. Data security technologies ("Privacy by Design") and Privacy Enhancing Technologies (PETs) give sound solutions to minimise the legitimate and reasonable privacy risks that a technology such as RFID might raise.

As there are, and will be, a multitude of technically different RFID applications adapted to the needs of the various economic sectors, the companies and sectors concerned should be allowed to develop and apply the adequate privacy by design and privacy enhancing technologies depending upon the legitimate and reasonable privacy risks related to the RFID application concerned.

> **Recommendation:** Self-regulation and guidelines enacted by industry should be welcomed

In addition to the legal provisions and the technical solutions, comprehensive industry codes of conduct or guidelines shall be encouraged for spreading best practices in order to comply with data protection rules and mitigate reasonable privacy risks. Already today, there are industry guidelines indicating how to inform consumers about the use of RFID technology and about the choices available to deactivate the tags. To be efficient, self-regulation needs to be coherently enforced by the industry: while admitting that by their own nature they cannot be legally enforced, self-regulations and guidelines would make little sense if they were not respected by their addressees.

5.2 Health and Environmental Effects

5.2.1 Health Effects

The expected increase in the number of RFID tags leads to the need to evaluate whether or not the current European legislation on the protection of health pertaining to electromagnetic fields suffices for the new scenario that will be created in the following years by RFID. This section aims to briefly refer to the legislation applicable in the field of health and the consequences that it has for RFID.

The issue of electromagnetic fields and public health is not new. Since the first apparatus emitting electrical radiation was introduced, several studies have warned against the harmful effects that overexposure to radiation has on human beings, and public authorities have taken adequate legal measures to protect citizens (such as those taken on mobile phones for example). In 1999, the European Union adopted a Recommendation aiming to limit general exposure to electromagnetic fields [Council Recommendation on the limitation of exposure of the general public to electromagnetic fields (0 Hz to 300 GHz) (1999/519/EC)]. This Recommendation sets up a number of restrictions and reference levels concerning the exposure to electromagnetic fields, with the purpose of ensuring a high level of protection for the general public. Furthermore, as certain workers are in daily contact with electrical and electromagnetic equipment, and are, therefore, more likely to suffer the consequences of their emissions, Directive 2004/40/EC on the minimum health and safety requirements regarding the exposure of workers to the risks arising from physical agents (electromagnetic fields) was adopted, so as to provide an adequate degree of protection at workplaces by establishing restrictions and requiring employers to take a number of measures.

In this regard, it can be considered that RFID falls within the scope of both legal texts, and that the potential effects on human health are, in consequence, covered by existing legislation. The real issue, however, will arise with the widespread development of RFID applications involving an increasing number of tags and readers.

To provide consumers with a comprehensive framework regarding health issues, they need to be reassured that RFID devices meet the requirements of all applicable EU legislation and that the applicable conformity assessment procedures have been applied. In order to give individuals confidence in RFID, the CE marking should, in principle, be used. CE stands for "Conformité Européenne" (European Conformity), and "symbolises the conformity of the product with the applicable Community requirements imposed on the manufacturer. The CE marking affixed to products is a declaration by the person responsible that the product conforms to all applicable Community provisions, and the appropriate conformity assessment procedures have been completed" (European Commission 2007).

However, using CE marking has proven to be difficult to apply when it comes to RFID. It does not seem feasible or helpful in this context, specifically due to the small size of the tags. Consequently, another solution may need to be considered in order to inform citizens of the technical compliance of RFID systems and health legislation. From this analysis, the following recommendations can be made:

> **Recommendation:** Monitor RFID developments to ensure compliance with EMF legal framework

As for all technologies using radio frequency, the current legal framework to protect citizens from over-exposure to electromagnetic radiation applies to RFID. The European Commission should continue to moderate a dialogue among relevant stakeholders to identify open points and engage in further legislation if deemed necessary.

> **Recommendation:** Other means different from CE marking to inform and protect citizens should be enacted

In this respect, an independent body in charge of inspecting RFID chips might be useful, either at a European or global level.

> **Recommendation:** Additional funds for technological research are welcome

The experience gained from first large live projects/applications of RFID is also an adequate method for addressing health concerns related to RFID.

5.2.2 Environmental Effects

As with any other technology, RFID raises a number of concerns related to its potential effects on the environment. The European Union has, however, enacted a consistent legal framework with the intention to cover all potentially harmful materials and avoid major effects on our natural surroundings. In this section we will analyse those laws and their application to RFID.

The most prominent issue with regard to environmental aspects of RFID usage is the question of recycling of transponders, products and packaging. With the widespread development of RFID applications, the components of RFID devices might cause significant problems to the environment when disposed of, since elements like silicon, nickel, copper or aluminium (all present in RFID devices) are contaminants for recyclers and manufacturers who use recycled material.

The European Union has adopted three fundamental legal texts to regulate environmental related matters of electrical and electronic equipment: Directive 2002/95/EC on the restriction of the use of certain hazardous substances in electrical and electronic equipment (so-called RoHS Directive), Directive 2002/96/EC on waste electrical and electronic equipment (so-called WEEE Directive) and Directive 94/62/EC on packaging and packaging waste. The objective of these directives is to restrict the use of certain components in the manufacturing of electronic and electrical equipment, as well as to improve the management of their disposal, trying to achieve a scheme as environmentally friendly as possible. Therefore, in principle, compliance with this legal framework ensures a high protection of the environment, and thus, all electrical and electronic equipment that are covered by the directives should not have limited effects on the natural surroundings.

It should then be established whether RFID falls into the scope of these directives. The wording of the WEEE Directive does not explicitly rule out that RFID chips could be seen as waste electrical and electronic equipment, but when the text was adopted, RFID had not yet reached its current stage of development. In the document "Frequently Asked Questions on Directive 2002/95/EC on the Restriction of the Use of certain Hazardous Substances in Electrical and Electronic Equipment (RoHS) and Directive 2002/96/EC on Waste Electrical and Electronic Equipment (WEEE)", the European Commission established a balanced approach that makes application of the WEEE Directive dependent upon whether RFID tags are placed on the packaging or in the equipment itself. Hence, apparently RFID cannot be considered as electrical or electronic waste itself, but is rather linked to the product it is attached to. As it stands, many of the circumstances in which RFID tags will actually be deployed would result in them falling outside of the scope of the WEEE Directive. In this respect, the European Commission is currently conducting a review of both the WEEE and RoHS Directives. One of the purposes is to clarify the scope of the WEEE Directive, so as to determine the application to RFID by formalising the criteria used in the analysed FAQ document.

Thus, at the present moment, no major challenges regarding RFID and recycling exist; nevertheless, new problems may arise once the time of mass deployment will come (Gliesche et al. 2008). Therefore, it is recommended to ensure continuous assessment of the environmental impacts together with the development of the technology and to monitor this alongside the appropriate legal framework. In this respect, it is advisable to conduct an:

Recommendation: Assessment of the environmental impact of RFID

The European Commission and other stakeholders have established a coherent position on how to treat RFID under WEEE and RoHS Directives. Nevertheless, continuous monitoring of technology development is advisable to make sure that once RFID has reached widespread item level use, the legal framework has been adapted if required. Therefore, the European Commission should support these monitoring activities carried out by academia and industry alike. Further discussion of the ongoing development of RFID technology could be required, and cooperation between industry and stakeholders would be useful in this respect (Gliesche et al. 2008). An early stage approach is considered as necessary. Examples such as the forthcoming study by the Institute for Future Studies and Technology Assessment in Berlin (commissioned by the German Federal Environment Agency) show how stakeholders (from industry to academia, from NGOs to government authorities) can join efforts in order to analyse the impact of RFID on the environment in the short- to medium-term. The study, entitled "Forecast of possible Impacts of RFID Mass Deployment on Consumer Goods for the Environment and Waste Disposal", is intended to be published in 2008, and will be the result of a comprehensive research process, which should be then discussed at the European level.

> **Recommendation:** Encourage research aiming at minimising environmental effects of RFID

Polymer technology could be used to produce tags without potentially harmful materials, such as copper or silicon, and energy-saving devices could be implemented.

5.3 Radio Spectrum

There is no doubt that technology plays a key role in the development of RFID. One of the essential features within the field of RFID is the availability of a harmonised frequency base throughout Europe. Although a number of instruments have been adopted in this respect, the evolving nature of the technology requires constant monitoring to ensure a coherent legal framework. This chapter aims to give a brief overview of the current European legal framework and the issues that radio spectrum might pose to RFID.

5.3.1 EC Legislation and other Policy Texts

The following table details the most prominent legislation within the field of radio spectrum:

Table 5.2 RFID-related EU radio spectrum legislation

Name	Description
Council Directive 87/372/EEC on the frequency bands to be reserved for the coordinated introduction of public pan-European cellular digital land-based mobile communications in the Community – to be repealed by Proposal COM/2007/0367 final – COD 2007/0126	Directive 87/372 required Member States to reserve a range of spectrum exclusively for GSM. Given the present situation, where a number of systems, such as RFID, have entered in operation it is necessary to repeal the directive and open those frequency bands to other devices.
Directive 2002/20/EC of the European Parliament and of the Council of 7 March 2002 on the authorisation of electronic communications networks and services (Authorisation Directive) (OJ L 108, 24.4.2002)	This directive was enacted with the purpose of harmonising and simplifying the rules and conditions for authorisation, to facilitate the provision of electronic communication networks and services.
2002/622/EC: Commission Decision of 26 July 2002 establishing a Radio Spectrum Policy Group (OJ L 198, 27.7.2002, p. 49)	Sets up an advisory group on radio spectrum policy in order to support the Commission on radio spectrum issues.
Decision No 676/2002/EC of the European Parliament and of the Council of 7 March 2002 on a regulatory framework for radio spectrum policy in the European Community (Radio Spectrum Decision) – (OJ L 108, 24.4.2002, p. 1)	Aimed to set up a legal framework in order to manage the growing demand for frequencies, by trying to harmonise them and systematise their use.
Commission Decision of 23 November 2006 on harmonisation of the radio spectrum for radio frequency identification (RFID) devices operating in the ultra high frequency (UHF) band (OJ L 329, 25.11.2006, p. 64)	Requires Member States to make available (within the period of six months after the entry into force of the Decision) the frequency bands for Radio Frequency Identification Devices on non-exclusive, non-interference and non-protected bases.
Commission Decision of 16 May 2007 on harmonised availability of information regarding spectrum use within the Community (OJ L 129, 17.5.2007, p. 67)	Intended to harmonise the available information on the use of radio spectrum within the Community, through the unification of both the format and the content of such information and by using a common information point, called EFIS.

5.3 Radio Spectrum

Name	Description
Communication from the Commission to the Council, the European Parliament, the European Economic and Social Committee and the Committee of the Regions: A market-based approach to spectrum management in the European Union COM(2005)400	As the digital dividend it is heavily fragmented, a European approach is necessary in order to fully reap the benefits from the switchover, Due to interference problems, standard digital broadcasting services and other communication services often cannot be operated in the same spectrum band. The Commission therefore suggests forming clusters of technologies using a similar type of communication network that can be clustered in closely related spectrum bands.
Communication from the Commission to the European Parliament, the Council, the European Economic and Social Committee and the Committee of the Regions: Reaping the full benefits of the digital dividend in Europe: A common approach to the use of the spectrum released by the digital switchover (COM(2007) 700 final)	Establishes the basis for the efficient distribution of the expected freed spectrum to happen with the digital switchover, by the end of 2012. Relevant for RFID since it suggests the possibility of relocating frequencies in order to improve the current applications.
European Parliament resolution: Towards a European policy on the radio spectrum 2006/2212(INI)	Establishes the basis for the efficient distribution of the expected freed spectrum to happen with the digital switchover, by the end of 2012. Relevant for RFID since it suggests the possibility of relocating frequencies in order to improve the current applications.

5.3.2 Analysis

Radio spectrum is defined as the entire spectrum of electromagnetic frequencies used for communications, with a rate of oscillation within the range of about 3 kHz to 300 GHz.

The availability of radio spectrum at adequate and affordable levels is a prerequisite for the generalisation of RFID. In Europe, there are two types of spectrum bands available to RFID. The first is exclusively reserved for it (limited number) and others are available on shared basis. At present, there are some problems concerning radio spectrum in the European Union, such as the lack of harmonisation and full availability of a "first generation" radio spectrum ranges. In addition, radio spectrum ranges are considerably smaller than the available frequency in non-EU countries, which hinders the technical development of RFID applications in the EU. In addition, there is a growing need for a pan-European program to tackle radio spectrum matters.

Although some important legal instruments have been recently adopted in order to address those issues, they are still far from the essential required organisation of the sharing of radio frequency. The Decision of 16 May 2007 which attempts to harmonise the availability of information on radio spectrum and the proposal for a directive repealing Council Directive 87/372/EEC, that intends to "free" frequency bands that are currently reserved for GSMs, might not suffice for shaping a consistent plan for the upcoming radio spectrum necessities.

With the switchover from analogue to digital communication services, additional radio spectrum will become available. This "digital dividend" is believed to offer a "public resource with an exceptional social, cultural, as well as economic potential" (Communication on a common approach to the use of the spectrum released by the digital switchover (COM(2007) 700 final). In the same document, the Commission has stated that a European approach on the switchover is needed.

Due to interference problems, standard digital broadcasting services and other communication services often cannot be operated in the same spectrum band.

The Commission therefore suggests forming clusters of technologies using a similar type of communication network that can be clustered in closely related spectrum bands. As the spectrum is fragmented, these clusters could only be formed following a common approach across the entire EU. More efficiency in spectrum management could also be achieved by establishing a common European spectrum plan.

Key requirements for the further implementation of RFID in Europe are a) the provision of suitable frequency bands and ranges and b) its efficient management. In 2005, the European Commission adopted a communication calling for a market-based approach in spectrum management. (COM(2005)400). It argued that the traditional approach in spectrum management of assigning individual spectrum rights and allocating the various bands to defined service categories "no longer seems appropriate for electronic communication services (…) Furthermore, technical development is making it less costly to enable devices to operate at various frequencies. The traditional model is not agile or responsive enough to enable society to reap the benefits of these developments. This leads to missed opportunities in terms of competitiveness, industrial development and jobs, innovation and choice of services for citizens."

Instead the European Commission suggested a market-based approach, planned to be established by 2010, to manage spectrum with more flexibility and efficiency. In order for the market-based mechanism to function properly, a substantial part of the spectrum should be put up for trading instead of a slow phasing in with one or several test bands. The new market based spectrum management scheme should follow the principles of technology and service neutrality.

For the European Commission, this market based approach could require further regulation in order to avoid unwanted effects such as monopoly pricing or a fragmentation into several non-useful slices of frequency spectrums.

Common to all proposals is a strong call for a better coordination of spectrum management at the community level – a position taken by the European Commission as part of its telecoms reform package in 2008.

Within this project an extensive list of the different frequency bands used for different RFID applications has been compiled (Walk et al. 2008). With regard to RFID, it has been criticised that the available spectrum for RFID in Europe is too narrow compared with other regions of the world, such as the United States of America or Japan, putting Europe at a competitive disadvantage. In addition, spectrum allocation is not consistent among the Member States.

In 2006, the Commission issued Commission Decision 2006/804/EC, according to which Member States should within six months of the publication of the decision, make available the 865–868 MHz UHF frequency sub-bands for RFID applications. It also establishes that manufacturers have to "ensure that RFID devices effectively use the radio frequency spectrum so as to avoid interference to other short-range devices" (Recital 3).

The decision has been a decisive step forward to create a level playing field for RFID applications across the Member States and begins to make up for some of the comparative disadvantages Europe has over the United States and other world regions. However, Member States have implemented this decision at different speeds, with France having been granted an exception of the decision, limiting the use of RFID application within certain distances around military installations (Commission Decision 2006/804/EC).

5.3.3 Conclusion

It is a fact that, in developed countries, there is virtually no "free" range of spectrum at the moment, although the expected digital switchover will free a wide amount. As afore mentioned, at the present time, it does not constitute a major problem for RFID, since the devices that are already working (UHF RFID) are adaptable to the existing frequencies. However, wider range of radio spectrum will be necessary to ensure massive development of RFID applications in Europe. Consequently, the following recommendation can be put forward:

> **Recommendation:** Ensuring an appropriate radio spectrum framework

The European Commission should keep up its commitment to harmonise spectrum management and allocation across the Member States. Additional UHF spectrum should be rendered available. As RFID and other emerging technologies rely on the availability of radio spectrum, harmonised frequencies and easy application procedures are essential. As there is a strong need for harmonisation, in consequence, the efforts of the European Union should be focused on harmonising the existing technical requirements for the development of RFID, rather than creating new ones.

5.4 The Intellectual Property Rights Framework

5.4.1 Policy Approaches

The right to be granted a patent on technical innovations or a copyright on a certain work is essential to promote progress. The European Policy Outlook RFID, drafted under the German Council Presidency in 2007 states: "The patenting and licensing of innovative technologies is an everyday process, and it is one of the prerequisites for technological progress. Companies want to get a return on investment made in their R&D efforts" (BMWi 2007). However, the framework has to strike the right balance of security, flexibility and fairness for both inventors and users of RFID technology. The question to which extent industry standards should rely on patents has to be dealt with in responsible way, balancing the interests of technology developers and manufacturers as well as users and society as a whole.

The European Patent Regime

Although not directly EU legislation, the European Patent Convention is one cornerstone of the European patent regime. Signed in 1973, it aimed to harmonise European patent procedures and created the European Patent Office (EPO). The work of EPO reduces costs and makes up for some of the disadvantages compared to US companies that, with only one application at the United States Patent and Trademark Office (USPTO), can obtain protection for their IPR for a far greater market than any of the national European markets. The enforcement of the European patent is done before national courts, making it difficult and in a lot of cases expensive to effectively enforce patent rights. In addition, a patent might have to be translated into several national languages, therefore creating extra costs. It seems, however, that some issues regarding translation have been resolved, as the so called "London Agreement" will enter into force in 2008 that simplifies translation requirements for patents filed with the EPO.

Compared to the US (still the main competitor in the RFID field), the European patent regime poses advantages and disadvantages alike. The fact that there still exists no single community patent certainly puts European companies at a disadvantage. One advantage of the European patent system in comparison to the USPTO's proceedings is the first-to-file approach. To be granted a patent in Europe, it is essential to be the first to file a patent application. The patent office of Japan employs the same rule. The USPTO, on the other hand, follows a first-to-invent approach, meaning that before a patent is granted, the applicant and the USPTO have to verify in a lengthy and expensive process, whether the device has been invented somewhere else.

In another respect, the US system has a definite advantage for those who wish to be granted a patent in the area of computer software and computer implemented inventions (CII). The USPTO regularly grants patents for computer software, regardless of its function or impact. In the EPO context, software products are generally exempt from patentability, (Art. 52 (2) (c) EPC). Only if the software products in questions are "susceptible of industrial application", "new" and "involve an inventive step" (Art. 52 (1) EPC), they can be patented. Whether or not computer software can or should be patented at all has been discussed for several years. One of the main lines of arguments runs between open source proponents and those that see software patents as a way to gain revenue for innovations, especially for SME (European Commission, 2000a). Attempts by the European Commission to harmonise software patent law across Member States by a directive introduced in February of 2002 (2002/0047/COD) were halted in the European Parliament in 2005, following years of heated controversy both among software companies and open source proponents as well among the European institutions. Some refer to the TRIPS (Agreement on Trade-related Aspects of Intellectual property rights, including trade in counterfeit goods) agreement within the World Trade Organisation (WTO) which states that "patents shall be available for any inventions, whether products or processes, in all fields of technology, provided that they are new, involve an inventive step and are capable of industrial application" (Art. 27 (1) TRIPS).

The Community Patent

Initiatives for a single Community Patent date back as early as the 1970s, when in 1975 nine members of the European Economic Community signed the "Luxembourg Community Patent Convention" which never entered into force. A second attempt in 1989 also failed because not all thirteen signatory states ratified the agreement.

After publishing a Green Paper on European patent issues in 1997, in 2000 the European Commission undertook a new attempt when it proposed the creation of a Community Patent (European Commission 2000b). The proposal called for the establishment of a Community Patent granted by the EPO. Patents were to be enforced by a patent tribunal established within the European Court of Justice (ECJ). However, the proposal was turned down by the Competitiveness Council in 2004.

In 2006, the European Commission started a public consultation on the Community Patent. One of the findings was that "industry, as well as other interest groups, generally supports the Community Patent as a way of addressing problems of the patent system." (European Commission 2006a). The same general opinion was voiced in a public hearing on July 12, 2006 (European Commission 2006b). Following the 2006 consultation, the European Commission issued a Communication on the European Patent System (COM(2007) 165) on April 3, 2007. The Communication highlights the correlation of patent activity and innovative power and calls for the creation of an internal market for patents. The Community Patent is still seen as the best overall approach, ending the current situation in which patent litigation in Europe is complicated and costly. According to a study cited in the communication (Van Pottelsberghe de la Potterie et. al. 2006), the cost for a patent application in Europe can easily be nine times higher than in Japan or the United States of America. As one of the latest developments, the Portuguese Presidency in 2007 stressed the importance of a quick solution to the patent issue as a prerequisite to fulfilling the goals of the Lisbon Strategy, and further actions in this respect have been announced by the Commission for 2008.

Related EU legal acts on Intellectual Property Rights

Directive 96/9/EC of the European Parliament and of the Council on the legal protection of databases

As RFID systems involve the creation of back-end databases, Directive 96/9/EC might in some cases also be relevant to gain legal certainty when implementing RFID. As a novel feature, the directive created a *sui generis* right for non-original databases (Art. 7 (1)). With this provision, it is intended to protect "the results of the financial and professional investment made in obtaining and collection of data and information". The database rights do not compromise any copyright that applies to the items of the database contents (Art. 3 (2)) In the evaluation of the Database Directive, the European Commission concluded that the level of copyright protection for databases, especially for non-original databases now covered by the directives sui generis provisions, had increased, also in comparison to the United States of America, but that this had "no proven impact on the production of databases" in Europe (European Commission 2005).

Council Directive 91/250/EEC of 14 May 1991 on the legal protection of computer programs

With Directive 91/250/EEC, the Council aimed at creating a common level of copyright protection for computer programs in all Member States. According to Art. 1, the purpose of the directive is to "protect computer programs, by copyright, as literary works within the meaning of the Berne Convention of the Protection of Literary and Artistic Works" (Recitals). The Berne Convention of the Protection of Literary and Artistic Works was first signed in 1886. It is today administered by the World Intellectual Property Organisation which is part of the

United Nations. The TRIPS agreement also incorporates the terms of the Berne Convention. Art. 5 (2) of 91/250/EEC allows for the making of back-up copies, Art. 5 (3) gives software users the right to correct errors in the software product. Art. 6 also gives the possibility for reverse engineering of software to ensure interoperability with existing systems. In a report on the implementation and effects of 91/250/ECC, the Commission concluded that "the objectives of the Directive have been achieved and the effects on the software industry are satisfactory (demonstrated for example by industry growth and decrease in software piracy)."

Directive 2001/29/EC of the European Parliament and of the Council of 22 May 2001 on the harmonisation of certain aspects of copyright and related rights in the information society (EU Copyright Directive (EUCD) or Information Society Directive (Infosoc))

The major impact of the Infosoc directive was to harmonise Member States legislation to counter "circumvention of any technological measures, which the person concerned carries out in the knowledge, or with reasonable grounds to know, that he or she is pursuing that objective" (Art. 6 (1)). This directive has an indirect influence on RFID, as RFID can be one technical measure to improve copyright protection and counter fraud and piracy, such as counterfeit technical components or brand name products.

5.4.2 Industry approaches

RFID Patent Pool

If standards rely partly or wholly on patented technologies, competition might be at stake. However, as previously mentioned, patents are an important means to ensure innovation and creativity in the development of new technologies. Thus, a balance has to be found in order to assure that the interests of both innovators as well as manufacturers and users of standard technology solutions are met. One way to moderate the impact patents may have in that progress is to form a patent pool. Patent pools might create an environment for a fair and non-discriminating competition by providing access to key patents for all market players. Currently, the most prominent patent pool in the IT and communication industry is MPEG LA. MPEG LA is the MPEG Licensing Administration that serves as a patent pool for the MPEG-2 standard. MPEG-2 is required for the production of DVDs. Currently MPEG LA is also preparing licensing programs for the evolving standards HD DVD and Blue-Ray Disc as well as for Digital Rights Management applications. Patent pools are a means to facilitate an easy licensing of patents as companies wishing to manufacture products or offer services for which patented technologies are needed only have to turn to one institution to receive a license for all patents needed to make a certain standard work.

In the RFID industry, "The RFID Consortium" was formed in the United States in 2005. It is largely modelled after the MPEG LA example. The following US companies have finally signed the contract: Alien Technology Corporation, Applied Wireless Identification Group, Inc. ("AWID"), Avery Dennison Corporation, Moore Wallace, an RR Donnelley Company, Symbol Technologies, Inc., ThingMagic, Inc., Tyco Fire & Security, and Zebra Technologies Corporation.

In 2006, the RFID Consortium issued a call for patents, inviting companies to identify patents they have been granted that are necessary for the practice of UHF RFID standards covered by EPCglobal Class Gen1 and Gen2 as well as ISO/IEC 18000-Part 6. As stated on its website, "all patents essential to the practice of the UHF RFID standards owned by participants in the licensing arrangement will be made available to interested companies via a single license on fair, reasonable and non-discriminatory terms." In November 2007, RFID Consortium LLC was established to facilitate RFID patent licensing following an obviously successful patent call the year before. The partners also include companies from Europe and Asia. Partners of RFID Consortium LLC are 3M Innovative Properties Co., France Telecom, Hewlett-Packard, LG Electronics, Motorola, ThingMagic, Inc., and Zebra Technologies. The program will be administered by Via Licensing, a subsidiary of Dolby Laboratories, Inc. In December 2007, RFID Consortium LLC submitted its business plan to the United States Department of Justice for a review of the legal and licensing requirements. This review also includes the list of essential patents that will be within the initial patent portfolio.

EPCglobal IPR policy

Another solution to the essential patents issue has been proposed by EPCglobal: companies wishing to participate in any of the Working Groups set up by EPC global in order to develop technical specifications for the development of RFID technology are asked to sign an IP Policy by which they engage themselves in offering the necessary patent claims royalty-free to the greatest extent. Necessary in this context means that "it is not possible to avoid infringement because there is no non-infringing alternative for implementing the Specification" (EPC global 2007). In its Intellectual Property Policy Working Group Declaration, EPCglobal states the purpose of the declaration is to "facilitate the adoption of such a set of Specifications while avoiding uncertainty to the extent possible regarding intellectual property claims in the Specifications. EPCglobal seeks to encourage the development, exploitation and competition of proprietary technology and innovative approaches to implementing such specifications, while avoiding blocking proprietary claims or monopolization of use of the Specifications". (EPCglobal 2003)

Section 3.1 of the EPCglobal Intellectual Property Policy Working Group Declaration states that those who sign it "shall grant to the extent that it owns or has a right to grant, a non-exclusive, non-transferable, non-sub licensable, worldwide royalty-free and otherwise reasonable and non-discriminatory licence" to other EPCglobal partners. With limiting its IP policy to necessary claims, EPCglobal hopes to encourage parties to "develop and benefit from exploiting proprietary implementations and improved systems and methods which utilise EPCglobal specifications". EPCglobal then go on to say that they "... encourage the development and use of intellectual property which is built upon a common set of interoperable standards." (EPCglobal 2007)

5.4.3 Open Source Approach: OpenPCD

Alongside industry-driven patent pools and royalty-free or RAND licensing schemes, the first open source solutions for RFID applications are beginning to enter the public discussion. One of the most prominent examples is OpenPCD, offering a "free hardware design for Proximity Coupling Devices (PCD) based on 13,56 MHz communication." (http://www.openpcd.org/). The overall goal is to provide people with technology helping them to detect and read RFID tags embedded in ePassports or other smart cards, following a mainly critical attitude towards RFID technology. The device is able to read cards supporting the ISO 14443, ISO 15693 and Mifare Classic standards. The hardware design has been released under a Creative Commons Attribution Share-Alike license, the necessary software under the name of "librfid" is released under GNU/General Public Licence (GPL). In February 2008, a version 0.2.0 of "librfid" was made public. The OpenPCD project originated among members of the Chaos Computer Club in Berlin, Germany. It aims at reading cards using the MRTD (Machine Readable Travel Document) standard issued by the International Civil Aviation Organisation (ICAO). After free software ("librfid" and "openMRTD") had been developed, a free hardware design was to follow – a process that has led to the OpenPCD specifications.

5.4.4 Conclusions

Patents are a normal feature in everyday business relations, and as such have positive and negative effects on technology development. A balance has to be struck in order to make patents a motor of innovation rather than a roadblock. It is also important to address the different aims of standards on the one hand and patents on the other, with standards being introduced to facilitate greater interoperability of systems across countries or industry sectors, and patents being granted to protect intellectual property rights for innovative inventions. Both are needed in order to spur further technological development and economic growth in Europe. If standards involve the use of patented technologies, then ways should be found so that the use of the standard will be possible for all those wishing to use the technology. Some methods of how this could be achieved (e.g. by patent pools) have been addressed in this section.

European companies are beginning to take an increasingly important role in industry initiatives, with Germany for instance being home to the second largest number of EPCglobal members after the USA, and with European and Asian companies now part of the RFID Consortium patent pool. The legal framework within Europe seems to be sufficient in order to promote further RFID technology development. Within the European Patent Convention, the implementation of the London Protocol as from May 2008 should make patent applications in Europe cheaper and less complicated. However, still a major impact can be expected from launching a genuine Community Patent. The latest initiatives in this respect will hopefully prove to be successful after decades of debate (Verheugen 2007). Therefore, it is recommended to:

Recommendation: Work towards a community patent

Continue collective political efforts to come closer to establishing a community patent for Europe, and take into account policy options for improving the current European patent systems (see also ETAG 2007).

Recommendation: Encourage European participation in patent pools

It is recommended to encourage and maintain European participation in patent pools and with standardisation initiatives and bodies, and also to establish regular consultations with US and Asian representatives.

Recommendation: Follow an international approach with regards to IPR

European Union should follow adopt an harmonized approach at the international level, using its weight of 27 Member States within the WTO (World Trade Organisation) and the WIPO (World Intellectual Property Organisation) to facilitate common IPR rules across the globe, as RFID even more so than other technologies, can only become successful on a global scale once the Internet of Things has become a reality.

Recommendation: Encourage an open approach to IPR

Encourage a lively and innovative IT sector by allowing for different approaches in how to manage IPR, from open source to patent pools to other industry agreements. Using RFID technology according to internationally accepted standards should be possible for as many users as possible at fair and reasonable costs.

> **Recommendation:** Continue the discussion on IPR and standards

Support Platforms such as the Global Interoperability Forum for Standards (GRIFS) launched under the 7th Framework Programme to foster dialogue with all stakeholders on patent, IPR and standardisation issues.

5.5 RFID Governance

As the Internet of Things (i.e. a global network of objects and corresponding information, that needs to be accessed anytime anywhere across the world, which falls outside the core of the scope of this study) is still a vision, a reliable estimate of the future scope of applications and the need for exchange between different industry branches is difficult to evaluate. From today's perspective, retail and goods manufacturers as well as logistics providers (corresponding to application fields Logistical Tracking & Tracing and Production, Monitoring and Maintenance in the RFID Reference Model) are likely to be among the main branches with the need to employ systems and standards that have to be accessed by large numbers of entities around the globe.

Other issues arise when novel applications come into focus. As every individual can have their own private website and use the internet to exchange data and information with other users, one day people might also want to exchange data and information with objects. This issue raises new questions, for instance regarding the storage of data of tagged personal objects in central or decentralised databases.

5.5.1 Observation of Current Debate on Internet Governance

The Working Group of Internet Governance (WGIG) within the World Summit of the Information Society (WSIS) held in 2003 in Geneva and in 2005 in Tunis has comprised a definition of internet governance, stating that "internet governance is the development and application by governments, the private sector and civil society, in their respective roles, of shared principles, norms, rules, decision making-procedures, and programmes that shape the evolution and the use of the Internet. It should be made clear however, that internet governance includes more than internet names and addresses, issues dealt with by the Internet Corporation for Assigned Names and Numbers (ICANN): it also includes other significant public policy issues, such as critical internet resources, the security and safety of the internet, and developmental aspects and issues pertaining to the use of the internet" (WGIG 2005). Members of the WGIG include international representatives from research, government, industry and civil society.

In its deliberations, the WGIG has identified several policy issues that should be addressed regarding internet governance. Among these are the unilateral control of the United States over the root zone file and system administration and the lack of formal relationships of government authorities with root server operators.

Further on, the WGIG suggests a division of roles and responsibilities amongst the three stakeholder groups, i.e., governments, private sector and civil society, that they should fulfil in the debate surrounding a future Internet governance model. According to the WGIG, *governments* should create an environment that favours ICT development, should develop, if applicable, laws, regulations and standards, should foster the exchange of best practices and engage in oversight functions. The *private sector* should promote industry self-regulation and the exchange of best practices, should develop policy proposals, guidelines and tools for policymakers and participate in national law making and foster innovation through its own research and development. *Civil society* should mobilise and engage in democratic and policy processes, engage in network building and bring in other views, e.g. grass-roots initiatives in order to facilitate a bottom-up, people-centred process. As a proposal for action, the WGIG concludes that "no single government should have a pre-eminent role in relation to international internet governance" and that all relevant stakeholders should be involved in a multilateral, democratic and transparent way. This call for an enhanced cooperation has also been stressed by the European Commission (European Commission 2006).

The current criticism is focusing on the role of the United States of America within internet governance. ICANN (the Internet Corporation for Assigned Names and Numbers) is currently overseen by the US National Telecommunications and Information Administration (NTIA). ICANN was originally formed to facilitate the transition of control over the internet's Domain Name System (DNS) from the US government to the global internet community. However, the US government's oversight over ICANN might soon come to an end. In its response to the midterm review of the Joint Project Agreement (JPA) with the US Department of Commerce, ICANN stated that the "JPA is no longer necessary" and leads to "a misperception that the DNS is managed and overseen on a daily basis by the US government" (ICANN 2008). Other governments are currently included by way of the ICANN Governmental Advisory Committee (GAC). The European Commission has acknowledged the positive role ICANN plays in internet management.

The role of the US government in DNS administration has been at the centre of numerous discussions. As a result of the aforementioned WSIS in Tunis and Geneva, the Internet Governance Forum (IGF) has been established as a global forum for discussions about alternatives in internet governance. In a resolution debated in January 2008, the European Parliament stressed the importance of the IGF as an international forum for the exchange on internet governance issues ensuring a democratic, transparent and multi-stakeholder dialogue (European Parliament 2008). In the parliamentary debate, suggestions were made for topics to be discussed at the next Internet Governance Forum meetings. These included broadening the internet governance debate to include issues such as the Internet of Things and the necessary frameworks for global RFID applications. However, the latest proposals for the New Delhi agenda do not explicitly mention RFID, but contain, however, several topics under which RFID-related aspects could be discussed (Internet Governance Forum 2008).

The other main point of criticism focuses on the role of the US-based company VeriSign as the main facilitator of the Internet Domain Name System and the administration of top-level domain names. Fears arise that the influence of one company might hinder Internet development as single economic interests might outweigh broader goals for an open internet environment.

5.5.2 *Legal Framework and Approaches to RFID Governance*

By its very nature, the internet has evolved over several years and has largely operated without a given legal framework. All over the world, lawmakers have closely followed the development of the internet in the past years and the importance it now plays in the everyday life of billions of people. Legislation has been passed in countries all over the world as well as within the European Union to ensure that the use of the internet does not conflict with national laws or international rights and conforms to norms and values predominant in the world's societies. Among the challenges tackled are, for instance, unwanted e-mails (spam), digital property rights in the internet age, restriction of certain internet content (e.g. child pornography) or enhanced consumer protection in online business transactions. However, the core of the internet, its governance structure, has not been subject to legislation, but to other forms of agreement (cf. for instance the Joint Project Agreement between the NTIA and ICANN). The same will be true for the Internet of Things. As it is still largely a vision with only some facets visible today, it has not yet been the subject of legislation, but rather of science and academia.

Current frameworks or models for RFID governance should not be mistaken to forestall Internet of Things governance models, but they might give a hint in which direction future discussions might lead. The most prominent example for a governance framework for RFID is EPCglobal. The EPC is a unique number identifier developed as a RFID data standard, mainly in retail environments. In 2003, EPCglobal was created as part of the GS1 organisation, ensuring a smooth transition from current barcode technologies and standards to future RFID application. EPCglobal is a membership-fee based non-profit organisation that, according to its own statement, is committed to open and royalty-free standards.

EPCglobal is the most prominent operator of an Object Name Service (ONS) – similar to the DNS used in the conventional internet. The ONS serves as a directory of EPC manager numbers and would be used by companies who would want to establish data with another company with whom it does not have an established relationship. It serves as a lookup service, providing a pointer to the information services provided by the manufacturer of the object. As an output the ONS produces a URL like known from the conventional internet. The URL then leads to an Electronic Product Code Information Service (EPCIS) repository that contains information on an individual Electronic Product Code. EPCIS would be implemented on manufacturer or company level. The standards for the EPCIS are still pending. The EPCglobal ONS is hosted by VeriSign – the US company mentioned above that also plays an important role in the internet's DNS. According to EPC global Inc. President Chris Adcock, there can be more than one ONS as long as they are interoperable (Adcock 2007).

EPCglobal is governed by a Board of Governors, in which different industry sectors are represented by global corporations from Europe, North America as well as Japan and China. For public sector applications, the United States Department of Defence (DoD) is represented. This has evoked criticism as some stakeholders fear a strong influence by the US Government on EPCglobal. This issue has been the topic of several debates in Europe. In his speech at the Lisbon conference "On RFID: Next Steps towards the Internet of Things", Bernard Benhamou, Professor at the Institut d'Etudes Politiques (Institute of Political Studies) in Paris, strongly warned against, for instance, repeating decisions made in the early days of the internet, which have led to the current criticism regarding internet governance. Instead, Benhamou called for the inclusion of all relevant stakeholders across all world regions. He said that different options for different countries and regions were necessary. Calls have also been voiced for more participation of different societal groups in the governance model of the Internet of Things – a position also summarised in the Commission Communication on steps towards a RFID policy framework (Communication on RFID in Europe: Steps Towards a Policy Framework, COM(2007) 96). The Communication gives reference to the Consultation process on RFID launched by the European Commission in 2006 in which 86 % of the respondents stated that the "system for registering and naming of identities in the future "Internet of Things" should be interoperable, open and non-discriminatory. […] It should not fall into the hands of particular interests that could use these databases and naming systems for their own ends, whether they relate to commercial, security or political aspects of governance". In addition, questions have been raised on the Security of the Object Name Service as suggested by EPCglobal (Fabian et al.).

Besides EPCglobal, other providers have begun to offer ONS services. Afilias, global provider of internet domain name registry services headquartered in Dublin, Ireland, has offered the Afilias Discovery Services (ADS) which is free of charge and compatible to EPCglobal standards. The ADS is based on the open Extensible Supply Chain Discovery Service (ESDS) protocol and ensures operability to other supply chain systems and business applications. The ADS is a first sign of an evolving market in the field of RFID governance that might lead to a fruitful competition of global information services (BRIDGE 2007). Most recently, GS1 France announced that it was planning the "nationwide implementation of the ONS root of the EPCglobal network architecture" in France, a project to become operational in spring of 2008 (GS1 France 2007). The ONS will be administered by Orange Business Services, a French company. In January 2008, first meetings took place by the EPC Network Committee that will bring together solution providers and network users to discuss the further evolvement of the French ONS

(GS1 France 2008). The French attempts could then lead to the development of an open governance mode. "Subsequently, this open governance model can be extended to incorporate various ONS systems from other parts of the world, both on technical and business aspects that would be administrated under a common set of rules. Drawing on the GS1 France project to initiate an ONS root in a European context", specific rules, naming standards, or security tools should be discussed and developed (Pauvre 2008).

The further development of RFID and subsequently the Internet of Things will most likely not be one mainly driven by legislation within the EU or other parts of the world, but rather an issue for a continued dialogue in appropriate forums and organisations. The planned discussions at the Internet Governance Forum in 2008 might be quite conclusive to get ahead in this important debate.

5.5.3 Conclusions

According to the Commission's Communication on "RFID in Europe: Steps Towards a Policy Framework" (COM(2007) 96), there are concerns about the openness and neutrality of the databases that will register the unique identifiers that lie at the heart of the RFID system, the storage and handling of the collected data, including its use by third parties. Hence, the issue of eGovernance is also perceived as a complex one: if the so-called Internet of Things is to be successfully accomplished, the data storage systems should be ethically and securely managed and the processes should remain interoperable and non-discriminatory.

As the Internet of Things is still a future vision, all stakeholders should take their time to discuss all issues properly and unbiased on the European as well as the international level. eGovernance structures should indeed be discussed in a broad scope. As a consequence of a fear of repeating decisions made when developing the internet, arguments have been brought forward to not have a centralised structure, but rather a decentralised one, also under the control of other countries or entities outside the US. The discussions should be open for all ideas and requirements. This means that the questions of how to store the data of a tagged item, how to provide the link between the item and the respective data, and how to manage the access to this data should be elaborated recognising legitimate demands for both open environments for ICT and internet development and companies' economic interest in developing new business models. The European Commission therefore should:

Recommendation: Keep a close dialogue with all relevant stakeholders

This dialogue should include all relevant stakeholders from industry, government, civil society and academia alike, as for instance represented in the RFID

Expert Group and the High Level Group on internet governance in order to foster an open, transparent multi-stakeholder dialogue ensuring interoperability, etc., and use international forums such as the Internet Governance Forum to discuss these issues and foster transatlantic dialogue with the United States on these topics. Take opportunities for exchange and proactively approach institutions such as EPCglobal to make sure that European interests are duly recognised.

> **Recommendation:** Encourage further scientific research in the field of RFID governance

Encourage further scientific research in this field, e.g. within the 7th Framework Programme. This research could focus on developing a decentralised structure of the EPCglobal Network with global governance participation as suggested in the European Policy Outlook RFID, which was finalised during the German EU-Presidency in June 2007, and could determine whether the EPCglobal model can serve as a pre-test for future Internet of Things structures.

Chapter 6
Technological Research Needs

6.1 General Technology Challenges

At a first glance there seems to be a contradiction between the hundreds of millions of already deployed RFID tags in car keys, contactless smart cards and library books (Harrop and Das 2006), and the increasing variety of industrial and academic R&D laboratories and projects working on RFID topics (see Wiebking et al. 2008 for an overview on international R&D activities). Yet, a closer look at the various application fields stated in Chap. 3 reveals the reason for the always increasing demand for further research and development activities. RFID is an umbrella term for many individual Auto-ID technologies with significantly different parameters and properties depending on the concrete application, as some examples may show.

For instance, most big retail players in Europe, such as Ahold, Carrefour, Delhaize, METRO Group, Rewe, and Tesco have put their focus on the logistical tracking and tracing of fast moving consumer goods. All of these companies are investigating RFID, are currently having ongoing pilots, or are starting limited rollouts in some specific product categories. In 2008, RFID pilots are moving towards real roll-out phases as part of normal logistics process control. Item-level tagging has already been realised with some of the most expensive products such as consumer electronics, apparels and accessories. The great challenge for broadly deployed item-level applications is the efficient and reliable mass manufacturing of the tags, as well as the integration of tags within the product packaging. On the other side of the supply chain, the food industry is searching for more secure and effective ways to track and trace food origins as well as temperatures during shipment and the quality of food in different logistic phases. So-called sensor tags (RFID tags combined with sensor technologies such as temperature measurement) might solve the problem or at least be part of a full solution which would enable better control tools for the food industry and, in the end, safer food for all European consumers.

Research and development tasks that arise here are quite different and focus on the low power consumption of sensors and computing capabilities on tag as well on new battery or energy harvesting concepts. In comparison to these applications the requirements for tagged legal documents are quite different. Passports, driving licences, legal contracts and shipping documents are increasingly equipped with RFID tags. In most of these applications passive, i.e. battery-less, tags will be sufficient. However, the tags have to be integrated in normal printing paper. Mechanical robustness therefore is an essential requirement. Using a typewriter for filling in additional information should not destroy a tag embedded on a document sheet. In addition to the unique identity of the tags in legal documents, their data content must also be secure against the attacks of hackers, be it identity theft, eavesdropping or forgery. There are also application areas such as health, public transportation or loyalty cards where Privacy Enhancing Technologies (PET) are needed, i.e. in applications where actual or potential risks to privacy should be reduced. For these services, PET may also offer, besides simple deactivation methods (see Chap. 5.1.3.5), more advanced methods like anonymity or pseudo-identity, unlinkability and unobservability (IPTS 2007). Thus in a number of cases tags, readers and backend-software must provide cryptographic functions which are adapted to the specific technological restrictions of RFID systems.

Besides these pragmatic requirements that stem from concrete applications there is also the long-term vision of the Internet of Things for which RFID may be a building block. Just like Mark Weiser stated in 1991 that in his opinion, the most profound technologies are those that weave themselves into everyday life until they are indistinguishable from it. Early forms of ubiquitous information and communication networks are evident in the widespread use of mobile phones, even more so than of the Internet.

Today, developments are rapidly underway to take this phenomenon further on in a two-folded manner. Firstly, short-range mobile transceivers will be embedded into a wide array of "smart objects", i.e. additional gadgets and everyday items, enabling new forms of communication between people and things, and between things themselves and secondly by adding sensing functions. Secondly, miniaturised sensors which are able to communicate with each other will build "sensor networks" that allow for a close and just-in-time monitoring of the real word. A new dimension will be added to the world of information and communication technologies. From anytime, anyplace connectivity for anyone, we will proceed to connectivity for anything. Connections will multiply and create an entirely new dynamic network of networks – an Internet of Things (Fleisch und Mattern 2005).

One of these varied technological applications, which make integration into everyday life possible and even encourage it, is RFID. In order to connect everyday objects and devices to large databases and networks, a reasonably simple system of item identification is important. Only then data can be collected, processed and items may be tracked in real-time. RFID offers this functionality and is therefore seen as one of the pivotal enablers of the Internet of Things. This leads to new additional requirements for future RFID systems. For instance, today's mostly centralised IT systems must be adapted to the decentralised decision mechanisms of the Internet of Things. Also, the safety, security and privacy of information will become even more important in this open and ubiquitous network.

In this chapter, the current technological barriers which apply to RFID will be identified and a technology roadmap will be presented that shows how to overcome these obstacles. Consequently, the chapter breaks down into two main parts. First, today's most prominent RFID technology challenges are discussed. Second, based on this analysis, the technology roadmap is developed which covers short-term, mid-term and long-term perspectives. The analysis of the bottlenecks and the specification of the roadmap is based on a detailed analysis of current R&D projects, two workshops with attendees from academia, RFID users, technology providers and system integrators and a final review by RFID experts. A detailed description of this process and further results can be found in (Wiebking et al. 2008).

6.2 Technology Requirements

Although many mature RFID systems have been in use for many years in certain applications (e.g. automation, production, access control, animal tracking, etc.) several technology bottlenecks still need to be addressed. Following the general design of RFID systems, these bottlenecks can be categorised as three main areas: tags, readers and the system as a whole.

6.2.1 Tags

The different types of tag technology – passive tags, semi-passive tags, and active tags – vary in their requirements, specifications and restrictions. However, some challenges apply to all three types of tags. These general requirements will be considered first before regarding the specifics of the different tag types.

- **Impacts of environment:** Adverse environmental conditions (e.g. the presence of liquids, metal) usually detune the tag antenna or coil, which results in a significant reduction of the reading range, or even a breakdown of the communication between tag and reader. Inter-tag interferences due to small tag-to-tag distance are still a problem in applications with high tag densities for item tagging (e.g. garments). Anti-collision technologies must be provided to address each tag individually. Improvement of reading rates in terms of completeness and speed is still a challenge, especially with big numbers of tags in one application.
- **Increased reading reliability:** An ongoing user demand is an identification rate of approximately 100% of the tag population. This is an important requirement for all multi-tag RFID systems (LF, HF, UHF). Actually some reading environments do not yield above a 95% identification rate, which is not acceptable for most applications.

- **Robustness in harsh environments:** One of the advantages of RFID technology is that it can operate in harsh environments (humid environments, high temperatures, dusty atmospheres, and mechanical impacts such as pressure or bending). There is demand for systems that can cope with even harsher environments. Silicon-based semiconductor chip technologies for consumer and industrial applications are usually limited to an ambient temperature range from −50°C up to approximately 150°C, which is still too restrictive for a number of applications.
- **Improvement of test coverage:** Individual testing of each tag of a production lot is an indispensable factor when striving for 100 % identification rates. Most current tests are done in near field, so reliable conclusions for far field operation are hardly achieved. There is a need for standardised methods for the measurement of product reliability, test coverage, and functional testing. Finding the relevant parameters is most important. Some customers ask for 100 % tested tags. Through modelling and computer simulation of the devices could be a chance to identify dependable criteria for functional tests.
- **Tag costs:** Especially for mass applications, many potential RFID users still see the total costs of tagging (tag cost plus mounting/packaging) as the essential factor of gaining enough return on investment (ROI). Therefore the whole manufacturing process from the raw materials up to the functional tag on the object − manufacturing of a chip, the contacting of the chip onto an antenna structure and the moulding of both components into a tag carrier or directly as an integral part of the product packaging − has to be optimised.
- **Environmental compatibility:** For future mass applications, where huge numbers of tags are used (item tagging) but where tags cannot be re-used, environmentally compatible or neutral RFID tags are needed in order to avoid electronic waste. Recyclable materials, material separation and (bio-) degradability are relevant topics.
- **Tags with display:** Optical identification systems (e.g. barcodes) can easily be displayed with static information (e.g. numbers, text, images). For some applications there is a strong need for displaying varying visible information to the user. With respect to RFID tags, this information could be related to the tag's data contents (ID number, object designator, etc.) or additional information (e.g. sensor data). Therefore it is important to develop RFID tags with integrated displays.
- **Nano power sensors:** The combination of RFID tags with sensor functions usually requires additional energy. The development of sensors with very low power consumption will be essential for the integration of more functions into tags. RFID sensors with high reading rates of their sensor parameters may not be implemented without additional energy sources. The higher the reading rate or the sensor update rate the higher the power consumption and the lower the battery lifetime. New concepts of "nano power sensors" with very low power consumption might be combined with active RFID tags in order to increase battery lifetime or the number of measurements per time period. A passive tag with low-energy sensor sourced from the reader field could be one of the goals for future development.

Passive RFID tags receive their power supply solely from the magnetic or electromagnetic field generated by the reader. Thus, the development of low power passive tags is an important goal.

- **Ultra low power tag ICs:** The maximum reading range of passive read-only tags mostly depends on two factors. Firstly, the right amount of energy that the tag needs for proper operation and secondly, a sufficient signal-to-noise ratio (SNR) must be present to read the tag's data with the reader. New technologies for lower power consumption on the tag side can increase the reading range.
- **Passive crypto tags:** For mass applications that require security functions, sophisticated cryptographic engines have to be developed and integrated into passive tags. Since these engines also must be powered by the magnetic or electromagnetic field of the reader, there is a trade-off between complexity of the crypto engine and the achievable reading range. As a matter of fact, there is a lack of tags with asymmetric cryptographic methods. This is caused by the fact that asymmetric cryptographic algorithms which provide a higher level of security than symmetric algorithms also need more computational power. The situation is even worse in the important UHF range. Since in this frequency range the energy consumption of the communication between tags and readers is higher than in the LF or HF range there is a general lack of cryptographic tags in UHF.

Semi-passive tags use an auxiliary energy source (in most cases a battery) to power the main and additional tag functions, e.g. clocks, temperature sensors, which need a continuous power supply whereas active tags use the auxiliary energy source also to actively generate radio frequency signals transmitted to the reader. These battery-supported tags have their own requirements:

- **New power supply concepts:** Battery-powered tags with mid range power consumption require new and more efficient battery concepts in order to cut down maintenance cycles and costs. This can be achieved by new battery concepts or even by power scavenging, i.e. the extraction of power from physical energy sources like temperature differences, pressure, vibration, light, magnetic fields, or electromagnetic fields.
- **Nano power wake-up receivers:** On active tags that receive and transmit RF signals either the tag has to be woken up by the reader, or the tag wakes up itself periodically. A significant amount of energy has to be used for the wake-up process. Most RF chips available on the market support the general reception mode only. Today, there is no very low power wake-up mode.

6.2.2 Readers

The following bottlenecks and R&D approaches relate to the readers used in RFID systems.

- **High Tag Identification Rates:** To achieve tag identification rates near 100 %, improvements are not only required on the tag side, but also on the reader side; for instance by additional signal processing of backscatter signals or improved antenna diversity schemes. In some long-range applications (UHF frequency range) mix-up reads of multiple tags often occur. Variable reading ranges, sector reading or new reading procedures – e.g. changing antenna lobes in the far field – in addition to the use of specific antenna arrays are promising approaches.
- **Multi-air-interface readers:** In some applications, tags with more than one air interface (e.g. HF and UHF) are used. Usually, readers are designed for operation on one specific air interface only. In order to read all tags with one reader within such applications (e.g. hand-held readers for manual checks), multi-air-interface readers would provide easier handling and cost reduction.
- **Miniaturisation:** The ongoing miniaturisation of communication devices in general raises a demand for RFID readers or reader modules in mini or even micro format (e.g. CompactFlash, SD-Card, MiniSD-Card format). The integration into mobile phones or PDAs, for example, would surely increase the usage and acceptance of RFID technology, as well as enable new applications.
- **Low-cost stationary readers:** Apart from the mini and micro RFID reader modules there is also a demand for less expensive stationary RFID systems. In order to achieve better coverage, e.g. at gate systems, multi-antenna and multi-reader arrangements might improve identification rates up to 100 %.
- **Reader-to-reader communication:** In order to provide higher reader densities (coverage) and better reader interoperability, the readers must be synchronised. New procedures and standards must be developed and introduced to reach this goal.

6.2.3 System Technology

Technology bottlenecks that address RFID systems as a whole rather than specifically reader or tag technology are presented in the following points.

- **Higher frequencies:** The current number of RFID systems and standards on higher frequencies ($\geq$ 2.45 GHz) is very small. On the other hand, there is a growing demand for smaller antennas, compact hand-held readers and modules as well as enhanced positioning resolution for localisation systems.
- **Ultra-wideband communication:** Most RFID systems provide data rates in the low or mid range sector. For applications requiring the transfer of large amounts of data in short periods ultra-wideband (UWB) communication concepts are welcome which use a large bandwidth of the frequency range to transmit the data in short-termed busts of communication at a low energy levels.
- **Real-time localisation and tracking:** Localisation is one add-on capability to RFID systems. Some applications do not only require the exchange of the ID number and some data of a tag, they also need the absolute or relative posi-

tion of the RFID tag. Although there are some active RFID systems with this functionality on the market, there is plenty of room for improvements in terms of positioning resolution, maximum tracking speed, reading reliability, etc.
- **High data rates:** The more functionality is packed into the tag chips, the more data has to be exchanged with the reader. Although today the requirements are primarily coming from e-government applications, in the future this might also be an issue for logistics and supply chain management.
- **New modulation formats:** Most RFID systems use load modulation or backscatter modulation in combination with basic modulation types (e.g. amplitude or phase modulation). The demand for higher data rates also raises questions for combined and more complex modulation schemes.
- **Improved system integration:** System enhancements and the introduction of new technologies should include all interfaces of the RFID system to RFID middleware systems. In order to provide second source capability, easy replacement and cost reduction, these interfaces should be standardised. Standardisation efforts should not only be limited to air interfaces and data structures but should also be extended to the reader/edge server interface.
- **Plug and play infrastructure:** Already implemented RFID systems often face a variety of different environments e.g. in terms of reflections and interference situations. Usually the system user cannot assess the actual environment and is unable to improve it for RFID usage. The RFID system itself should be able to adapt to various environments in order to approximate such a thing as a plug and play RFID infrastructure.
- **Increased interference immunity:** Many RFID systems can easily be interfered with simple jamming devices or other RF equipment that use the same frequency allocation. Such devices might even be legally operated in some countries. There is a need to identify such problems and attempt to ban these devices or to develop techniques to easily detect such devices, or render them ineffective. This is an extremely important topic as there are currently no satisfactory solutions in sight.
- **Security, data protection, privacy:** Many present and future RFID applications – including the envisioned Internet of Things – require safe and secure data transfer and data storage on the RFID tag. Research and development is needed, in order to prevent tag cloning, the unauthorised access to (private) data, eavesdropping, man-in-the-middle attacks, etc. effectively. Security and privacy enhancing technologies like secure authentication, secure authorisation and data encryption use cryptographic procedures. Most of today's RFID systems use proprietary symmetrical cryptography whereas there is a strong need for more secure asymmetrical or even more advanced cryptography.

6.3 RFID Technology Roadmap

RFID is not just radio technology but a cross-over technology which comprises microelectronics, micro systems, software technology as well as cryptology. Several RFID roadmaps have recently tried to cope with this variety, most prominent the European Technology Platform on Smart Systems Integration "EPoSS" (EPoSS 2007) and the Cluster of European RFID Projects "CERP" (CERP 2007). The following roadmap has been elaborated in close cooperation with both activities but goes more into detail. Other RFID studies like "BSI 2005", "IPTS 2007" and "Bizer et al. 2006" put their focus on other topics and provide only less detailed roadmaps. "BMBF 2007a" is a valuable source for technology issues regarding RFID and IT security.

Besides these roadmaps, which are driven mainly by technology providers and academia, there is the user-driven organisation Global Commerce Initiative which has defined six tag classes in its 'EPC Roadmap' (GCI 2003). From class 0 up to class 5 the RFID system performance and capabilities of the tags increase (see table 1). Class 0 tags are simple passive tags with a fixed number whereas class 5 tags stand for the fully developed concept of a sensor network. Only class 0 and class 1 are accompanied by corresponding EPCglobal standards at present. This classification system is used mainly by the EPCglobal community. The currently most advanced tag technology is the so-called Class 1 Generation 2 (short Gen2) technology, describing passive read/write tags at UHF (840–960 MHz). Specifications for tag classes 2 to 5 (semi-passive to active tags) still have to be defined (GCI 2003, p. 16–17).

In addition, the often cited EPC Roadmap may be seen as the common understanding of the RFID community concerning in which direction future systems will evolve, but the roadmap itself is not very speeific on technological aspects. The roadmap developed within this project intends to fill this gap by focussing on the technology bottlenecks from the previous subsection.

Table 6.1 Tag classes of the EPC Roadmap

EPC tag	class	Tag type/tag class capabilities
Class 0	passive	Read only, i.e. 64 bit EPC number hard coded by manufacturer
Class 1		Read, write once, i.e., EPC number with 96–256 bits can be encoded onto the tag later in the field
Class 2		Read/write: larger memory, user data
Class 3	semi-passive	Class 2 capabilities plus a power source to provide increased range and/or advanced functionality, e.g. sensors
Class 4	active	Class 3 capabilities plus active communication and inter-tag communication
Class 5		Class 4 capabilities plus communication with passive tags as well

6.3.1 Packaging

At present most tags are based on silicon technology. The chip is mounted onto a substrate and connected to the antenna, typically by pick and place of the chip. This causes around half of the tag costs. New concepts for mounting tags onto objects and integrating them into packaging may help to reduce costs and help to improve tag reliability and durability. Technologies like transfer moulding of tag chip plus tag antenna into plastic boxes, direct printing of antennae on flexible substrates (textiles etc.), using the product as a substrate, may help to further reduce costs of tagging (chip-in-paper, chip-in-foil, chip-in-substrate).

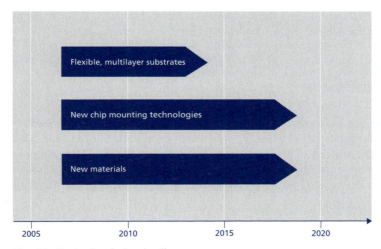

Fig. 6.1 Packaging design timeline

6.3.2 Chip Design

Current active tags often use conventional RF, analogue and digital components. Since active tags commonly use full RF transceiver chips, they are able to establish further communication links in addition to the RFID link. A cost effective solution might require a monolithic/single chip device to be integrated into an RFID tag, microcontroller, and additional RF components (e.g. for Bluetooth, ZigBee, Wireless LAN, FM functionality). Further miniaturisation and integration of active RFID tag functions into single chip designs may help to reduce the currently high costs of active tags significantly. Many currently used readers do not use highly integrated technology. For LF and HF there are integrated circuits available, which include major reader functions. Most of the UHF readers use conventional discrete components, especially in the RF section, which makes the reader expensive. Designing reader chips (including UHF) is a challenging task for reducing reader costs and reader size.

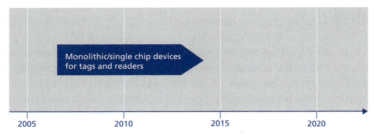

Fig. 6.2 Chip design timeline

Low-cost mini and micro readers enable new applications, e.g. integration of reader functionality into communication devices as PDAs, mobile phones, notebooks, navigational systems, etc. Chips may also be designed for low power consumption.

6.3.3 Energy Aspects

Power consumption is a great challenge for RFID tags. The maximum reading range of passive tags mostly depends on how much energy the tag can send back to the reader. The amount of energy the tag needs for proper operation therefore reduces the available energy for answering. New technologies for lower power consumption on the tag side can enlarge the reading range and improve the read-

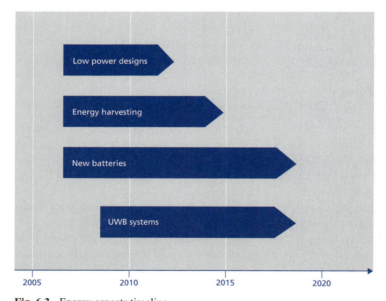

Fig. 6.3 Energy aspects timeline

ing reliability. In order to increase the reading range, the development of new ultra low power designs for RFID tag circuits is desirable. Lower power consumption for the tag's main circuit also clears the way for additional tag functionality (sensors, crypto engines, data processing, etc.).

Power scavenging (energy harvesting) technologies, that convert energy out of physical energy sources (e.g. temperature difference, pressure, vibration, light, magnetic field, electromagnetic field), may help to replace conventional batteries. The combination of RFID tags with sensor functions usually requires an additional energy source, in most cases provided by semi-passive or active tags. New battery technologies (e.g. polymer batteries, fuel-cells, paper batteries) support increasing functionality or longer battery lifetime. The development of special receiving modes (nano power wake-up) together with high sensitivity will also help to extend battery lifetime.

An interesting approach for low (nano) power communication with high bandwidth efficiency is ultra-wideband (UWB) technology. This technology might be used to introduce new RFID systems that require high data rates together with moderate reading ranges.

6.3.4 RF Technology

Ideally, tags should work under all environmental conditions. When attached to metal or near liquids especially, the reading range of tags may be significantly

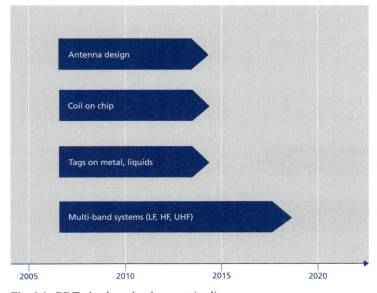

Fig. 6.4 RF Technology development timeline

reduced. When many tags have to be read simultaneously, inter-tag interference may occur, especially for UHF tags. These issues can be resolved or at least improved by optimised antenna design and by tags supporting multiple antenna connections.

For a short-range communication there is no need for a large external antenna. Instead, the antenna may be part of the chip itself ('coil on chip'). First prototypes of this technology are available but there is still need for improvement.

Finally, there is a need for RFID tags that support more than one air interface (e.g. HF + UHF, LF + HF). New technologies might improve the reading ranges and performance over all allocated frequency bands (multi-band, multi-port, multi-antenna, multi-polarisation). This is one of the major business challenges (e.g. global supply chain management).

6.3.5 Manufacturing

Low cost high quality tags are essential for potential RFID users to gain enough return on investment (ROI). Currently, low-cost silicon-based RFID tags are about € 0.10 at higher quantities, a price goal of € 0.05 or less per RFID tag (chip + antenna + substrate) is still a challenge (IdTechEx 2007).

The classic RFID tag manufacturing process comprises the manufacturing of a chip, the integration of the chip into an antenna structure and the moulding of both components into a tag carrier or directly into the product packaging. In order to reduce the number of production steps and simultaneously the tag costs, new manufacturing methods have to be developed.

An example of process optimisation for silicon based processes has been introduced by Alien Technology (USA). They claim that using optimised wafer processing by a fluidic self assembly of the chips they can achieve a 15% to 20% higher yield and about 200 times higher assembly rates than conventional methods (Alien 2008).

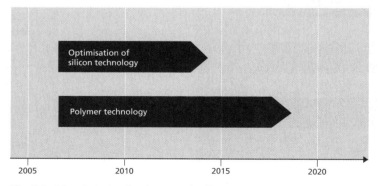

Fig. 6.5 Manufacturing development timeline

6.3.6 Systems

Ultra-wideband technology and other high speed data communication methods can help to resolve the growing demand for high data transfer rates. Most RFID systems use load modulation or backscatter modulation in combination with basic modulation types (e.g. amplitude or phase modulation). The demand for higher data rates could be met by combined and more complex modulation schemes, like quadrature amplitude modulation (QAM), and quadrature phase shift keying (QPSK).

The current number of RFID systems and standards on higher frequencies (≥ 2.45 GHz) is very small. On the other hand there is a growing demand for smaller antennas, compact hand-held readers and modules as well as enhanced positioning resolution for localisation systems. The usage of higher frequencies is one solution to these demands. A proposed air interface standard at 5.8 GHz has been withdrawn. New standardisation attempts for this and higher bands would be helpful.

Standardisation efforts should not only be limited to air interfaces and data structures but should also be extended to the reader/edge server interface. Interoperability, second source, easy maintenance and replacement, as well as cost reductions beyond the tag/reader system will be the consequences.

Implemented RFID systems often face a variety of different environments. Usually the system user cannot assess the actual environment and is unable to improve it for RFID usage. The RFID system itself should be able to adapt to various environments in order to approximate such a thing as a plug and play RFID infrastructure. Modelling and simulation software and systems are very slow, expensive, and limited to certain system parts. Reliable simulation software, capable of simulating an entire RFID system would greatly support implementation processes.

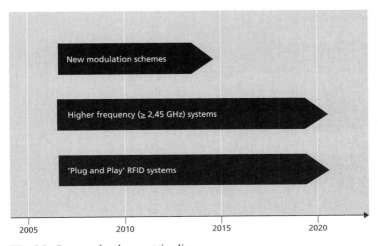

Fig. 6.6 Systems development timeline

Many RFID systems can easily be interfered with by RF equipment that uses the same frequency allocation (e.g. small range devices). There is a need to identify such problems and suggest that such devices are banned, or to develop techniques to easily detect such devices and render them ineffective.

6.3.7 Readers

Usually readers are designed for operation on one specific frequency. There is a demand for multi-frequency, multi-standard readers for some applications. Readers that can operate on several LF, HF and UHF standards simultaneously may recognise many kinds of RFID tags (ISO, EPC). Reader/edge server interface and parts of the control logic may be used in common for several air interfaces, reducing total cost of reader and RFID middleware. The ongoing miniaturisation of communication devices in general raises a demand for RFID readers or reader modules in mini or even micro format (e.g. CompactFlash, SD-Card, MiniSD-Card format). The integration of reader technology into mobile phones will surely increase the usage and acceptance of RFID technology, as well as enable new applications.

New intelligent reader concepts will not only reduce the size but also the cost of the readers to enable large-scale integration in small communication devices. Multi-antenna and multi-reader arrangements might improve identification rates and achieve better coverage, e.g. at gate systems. Higher reader densities and better interoperability require synchronised readers. New procedures and stand-

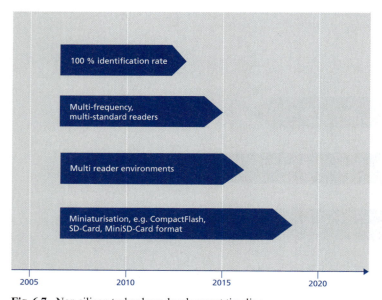

Fig. 6.7 Non silicon technology development timeline

ards must be developed and introduced to reach this goal. Reader networks may also support inter-tag communication.

In some applications, a large number of tags within range of a reader must be completely identified in a short period of time (e.g. single items on a palette). In this situation, reading rates are still a challenge. To achieve tag identification rates near 100 % improvements are not only required on the tag side, but also on the reader side. Anti-collision technology, antenna technology and signal processing may help to reach this goal. Variable reading ranges or sector reading would be useful features to reduce multi-tag mix-up reads.

6.3.8 Non-Silicon Technologies

Surface Acoustic Wave Devices (SAW)

Silicon-based semiconductor chip technology is usually limited to the temperature range from –40°C up to approx. 200°C. Surface acoustic wave (SAW) technology is not based on silicon substrates but on materials like lithium niobate or lithium tantalate. It extends the operating conditions for tags into more robust environments. The SAW approach offers a number of remarkable capabilities that cannot be achieved by conventional RFID systems: a high temperature resistance up to 400°C, readable at high velocity, high data rates up 1000 queries per second, and long reading range up 10 m even for fully passive tags. Additionally, SAW provides inherent sensor features for parameters like temperature, pressure, and strain gauge.

Polymer Electronics

One way to further, simple mass production and therefore cost reduction is to use new non-silicon based technologies like polymer (organic) electronics and reel-to-reel manufacturing procedures. The antenna and circuitry can be printed and integrated into the packaging with simple processes.

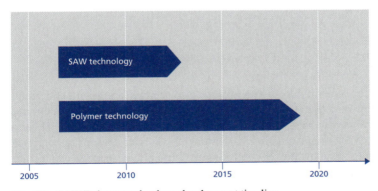

Fig. 6.8 SAW/Polymer technology development timeline

Another approach to manufacturing cheaper tags is the polymer technology. Costs for mass production of organic tags are estimated around €0.01 per tag. By printing tags (chip and/or antenna) using organic semiconductors, the manufacturing process is simplified significantly. Companies like PolyIC, ORFID, OrganicID, and NXP are pursuing research in this direction. PolyIC showed a first printed HF tag (13.56 MHz) in 2007. Even though first products will be realised soon, researchers estimate that it will take several years (up to 2015) until mass production capability for these tags is achieved (Harrop and Das 2006). According to current knowledge the highest frequencies achievable for organic tags in the next few years will be 13.56 MHz. In a later stage also the UHF range might be reachable.

6.3.9 Bi-stable Displays

For some applications there is a need for displaying varying visible information to the user (e.g. postal services, retail). RFID tags can be combined with visual displays, showing tag-related data. Currently available liquid crystal displays (LCDs) are too energy-consuming for low power (passive) tags. Bi-stable nano power displays only need power when their content changes. By showing the tag information, the display may help to further acceptance of RFID technology by the end user.

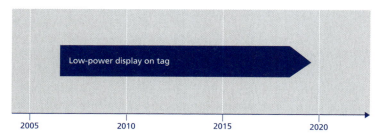

Fig. 6.9 Bi-stable display development timeline

6.3.10 Sensors

Besides the ID functionality some applications also require to measure and monitor physical parameters (e.g. temperature, pressure, humidity). A sensor, integrated in the tag, can be used to send the measured data to the reader via the RFID communication link. Power consumption is a critical issue with sensor/tag combinations. Sensors with very low power consumption (nano power sensors) can also be integrated into passive tags. In this case the energy of the reader field has to supply the complete RFID tag including the sensor.

6.3 RFID Technology Roadmap

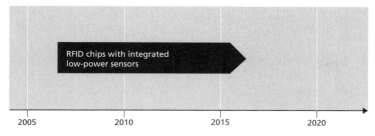

Fig. 6.10 Sensor tags development timeline

For the extension of passive RFID tags with sensor functions, sensors that use a bi-stable effect or that do not need energy from the tag can be used as no-power sensors.

6.3.11 Cryptography

For applications that require security functions, sophisticated crypto engines, especially with asymmetric approaches, have to be integrated into the tag. Since these engines must be powered by the available energy of the tag, there is a trade-off between complexity of the crypto engine and the other functions of the tag (i.e. reading range). The combination of new nano power technologies and new optimised crypto engines may pave the way to passive crypto tags with acceptable reading ranges and crypto performance. In general, there are two ways to cope with the hardware restrictions of RFID: either by more efficient high-performance algorithms or by radically simplified crypto engines.

An important research approach for advanced crypto engines is the use of elliptical cryptological functions (instead of today's prime factors). Low-power or light-weighted crypto engines can be based on minimalistic cryptography ap-

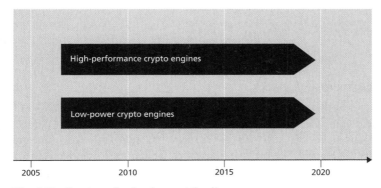

Fig. 6.11 Cryptography development timeline

proaches which move all security functions from the tag to the reader or the backend systems or on physical one way functions and physical non-replicable functions.

Finally, identity management, which is a corner stone for privacy enhancing technologies and which is today mostly based on public key infrastructures, must be adapted to the huge numbers of tags in future RFID mass applications.

6.3.12 ICT Architectures

RFID applications that need networking functionality (e.g. sensor networks, ambient intelligence systems) require tags with the ability to communicate with other devices e.g. tags, nodes, sensors, routers, hubs (ICT, inter-tag communication). Active tags can use their increased communication capabilities to connect to networks or to build networks (e.g. mesh networks). Smart tag networking, inter-tag communication (ad-hoc) and in general wireless (sensor) networks may be established. Such networks help to achieve higher reading rates (time and completeness) and/or higher reading ranges (multi-hop reads). RFID tag systems may interoperate with other networks (e.g. low power sensor networks based on ZigBee technology).

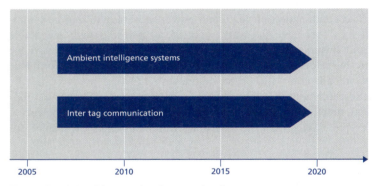

Fig. 6.12 ICT architecture development timeline

6.3.13 Environmental Aspects

RFID systems that are used in the retail sector or otherwise come into contact with consumers need to be analysed to ensure that relevant environmental aspects have been discussed. For many applications where huge numbers of tags are used (item tagging) and the tags cannot be re-used, environmentally compat-

6.3 RFID Technology Roadmap

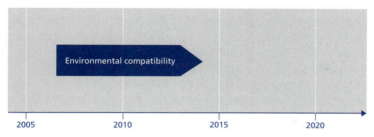

Fig. 6.13 Environmental research timeline

ible or neutral RFID tags are needed in order to avoid electronic waste. The introduction of non-toxic electronics (e.g. polymer) may help to create environmentally compatible tags.

Chapter 7
R&D Environment

RFID is a key technology in the fields of ICT for the European industry. Although no longer an emerging technology, the breadth of its potential application and the number of open issues on their way to full implementation and market acceptance still require substantial R&D support in many domains. A European research policy is an important enabler to realise the full potential of RFID.

When setting up a research policy, a decision about the "what" and the "how" is essential i.e. which topics should be supported and what the support should look like. In addition to the content of technological and societal research, and the creation of tools, measures and infrastructure to implement, another important issue is having the means to transfer research findings into economic benefits for the widest possible selection of stakeholders, e.g. companies, customers, consumers, governments etc.

There is a general impression amongst the players in the RFID value chain that programmes and measures to support RFID are few and not always aiming in a supportive direction. Therefore, in this chapter we assess the current RFID R&D policy in Europe via its present implementation, the European and national RFID R&D programmes, and compare this current status with the requirements as perceived by the providers and users of this technology, and make recommendations as to the content and the way of implementation of an optimised European RFID R&D policy.

7.1 Outline and Approach

Broadly speaking, the recommendations on a European RFID R&D policy are based on the analysis of the current status of R&D programmes and projects, starting with a survey of the present national subsidy programme lines of key countries and of transnational programmes of EUREKA and the EU (EUREKA is a European research initiative aiming at cross-border cooperation between industry and technological research institutions, www.eureka.be). Alongside this analysis, and taking into account the content of the subsidy programmes and projects analysed, ideas and proposals for specific R&D topics were collected via interviews in this research project, and via literature research (IPTS 2007, Van de Voort and Ligtvoet 2006, European Parliament STOA 2006, Buckley 2006, SEC(2007)312, BMWi 2007). Only documents dated between 2000 and June 2007 were used in the analysis.

The assessment and analysis were carried out against a set of criteria, outlined in the following section. Survey and analysis of current R&D programmes are covered in the following chapter and we want to give suggestions as to:

- Which R&D topics should be addressed with priority?
- How the EU could optimise the return on subsidy money spent (both national and European).
- In which way the EU could act as sponsor, enabler, or lead customer to support the adoption of RFID in areas of public interest.

7.1.1 Assessment Criteria of R&D Support Programmes

In order to have a consistent method of work, we first had to define criteria to assess the present subsidy programme and projects. These criteria divide broadly into two categories. Firstly, general programme characteristics were analysed, such as:

- Do programmes exist?
- Are they accessible for all potential participants (large companies, SMEs, academic institutions) or do they focus on certain sectors (e.g. SMEs)?
- What is the probability of a successful application?
- Does the programme support single companies or are consortia required?
- Are these thematic programmes/call systems or are they open for any topic?
- Does it cover all phases of innovation all the way to market introduction, or are there restrictions (e.g. only applied research etc.)?

Specific criteria relating to needs and necessities for RFID were also used in the analysis. One of the most important points which had to be addressed was establishing whether the programme and subsequent projects addressed the main bottlenecks which hinder the application of RFID. In order to establish this, a brief analysis of these bottlenecks had to be carried out.

The main inputs for these specific criteria were the publications which were collected via desk research (IPTS 2007, Van de Voort, Ligtvoet 2006, BMWi 2007, Bovenschulte 2007). For more information on these documents, please refer to the study of Pavlik and Hedtke (2008). These publications also proved to be a valuable source of information on perceived bottlenecks (or challenges) that had to be addressed to achieve a profitable implementation and a broad market acceptance of RFID technology. By taking a cross-section of opinions voiced in the literature, as to which challenges have to be addressed to come to a full implementation of RFID in the various use cases, it seems that there is general agreement on the following points:

a) Technological challenges in the tag/reader system, amongst others the required additional functionality for novel applications (e.g. smart systems), the cost issue, and reliability and dependability issue necessitating a case-by-case optimisation of the application.
b) The lack of business models, mainly for the SMEs in the RFID value chain, whose access to using RFID is hampered by the unavailability of generic architectures (building blocks) and lack of a fair sharing of costs and benefits in the value chain.
c) The reluctance of the consumer to embrace the technology and its services due to unsolved or inadequately addressed data security and privacy issues as well as patchy information and customer involvement policy.
d) Lack of uniform standards and guidelines across Europe and across the applications.

For our analysis we retained the issues a) to c) with a focus in c) mainly on the technological aspects of privacy by design. The standardisation issue is covered in Chap. 3 of this book, the guidelines issue in Chap. 4.

7.1.2 Methodology used for the Analysis

Initially, generic programme characteristics (see above) were investigated primarily via using available information sources (publications, homepages of funding agencies and ministries etc.), combined with interviews of responsible people in companies participating in relevant funding schemes and also personal contacts with representatives of European programme lines have been used to collect information. This worked particularly well for the European programmes; however information on certain programs at national level was quite difficult to collect, due to confidentiality rules which national funding agencies have to adhere to.

For the specific RFID-content related information we used a mapping technique based on the RFID Reference Model to determine the application focus of the individual programme or project. The model could then be extended with the following additional criteria, based on the issues a) through c) outlined above.

- Coverage of tag and reader technology aspects
- Coverage of systems aspects
- Coverage of business model and ROI aspects
- Coverage of privacy and data security aspects

An analysis was then carried out to establish to what extent the programmes covered the major challenges as perceived by the representatives of the RFID value chain. This was made for each programme line of the European projects, and also for each country when analysing the national programmes.

The results of this analysis are given in graphical form in each individual chapter covering specific countries or European R&D support programmes.

7.1.3 Programmes and Countries Considered

This section gives an overview of the RFID-related funding programmes, both at a national and European level, in the period from 2000 to early 2007. This overview is certainly not exhaustive or complete, as access to programme and project information on national level was sometimes not easy to obtain. We do believe however, that from the database created, conclusions and recommendations on EU R&D policy can be drawn.

All programmes supporting RFID and RFID-related technology and innovation were considered, from research to market introduction, and also programmes generating and supporting innovative infrastructures and cooperation models beneficial to RFID. Only Programmes were considered where we found proof of RFID-related activities and projects being executed in the context of these subsidy programmes. The following programme types were included in our assessment:

- All national programmes in the selected countries that either directly support RFID technology and application or support infrastructural measures with significance for RFID
- European programmes with the same target setting; these include
- The EU Framework programmes FP5 and 6,
- EU regional programmes (e.g. Interreg)
- The EUREKA Programmes (Clusters, Umbrellas and bottom-up projects)

For the selection of countries for this analysis we had several restrictions due to availability of information. We focused on the main European players based on the general activity level in RFID and the number of real business cases reported, as concluded from articles in the most important technical journals dedicated to RFID. Further we checked to the greatest extent possible, whether smaller countries or new Member States had a comparable activity level (as a means to measure the importance those countries attribute to RFID as a perceived core technology for Europe). The decision for analysing the research activities in these countries originates to a great extent to the availability of information. As stated above it was not possible to gather information for every European country. As a heuristic approach we decided to use the available information because drawing conclusions from this sample is still better then deciding without a sound basis. The candidates for the analysis actually come from three distinct groups:

1. Germany, France, UK, Italy, the Netherlands and Spain as representatives of large EU Member States with high visibility of the RFID topic, judged by the number of publications, projects and use cases
2. Austria and Finland, as representatives of the smaller Member States spending a relative high percentage of the GDP on R&D
3. Poland, the Czech Republic and Hungary as representatives of the central European Member States

For these countries a systematic overview was created, starting from the structure and content of the national subsidy programmes and then focusing on programme lines which support RFID including examples. We further analysed the focus and impact of those programmes using the evaluation criteria outlined above.

7.2 Analysis of National Programmes

In this section we will give a brief summary per country for national R&D programmes related to RFID. For more details please refer to (Pavlik and Hedtke 2008). Below is an overview of the countries selected and their respective R&D programmes.

Table 7.1 Overview of national programmes

	Funding agencies found involved in RFID R&D	Number of dedicated RFID program lines found	Number of RFID-related projects analysed
Large Member States			
Germany	6	2	16
France	4	1	17
UK	1	none	1
The Netherlands	2	none	2
Italy	none	none	1
Spain	2	none	4
Small Member States			
Austria	3	1	17
Finland	1	none	10

7.2.1 Germany

Funding for R&D projects on national level mainly come from the following four organisations:

- BMBF (Bundesministerium für Bildung und Forschung): German Federal Ministry of Education and Research
- BMWi (Bundesministerium für Wirtschaft und Technologie): German Federal Ministry of Economics and Technology
- DFG (Deutsche Forschungsgesellschaft): German Research Foundation
- FhG (Fraunhofer Gesellschaft): Fraunhofer Society

The search for current and past RFID Projects and programmes gave the following results:

BMBF Programmes

Project Call "Smart Labels in Logistics": This call was launched in 2004/2005 as part of the "Mikrosystemtechnik" (microsystems, MEMS) programme, and specifies the programme focus as "support for industrial cooperative projects which deal with open issues of micro systems technology in smart labels/transponders and mobile readers, and for projects which want to test these smart label prototypes in field application tests". A total of 11 projects were identified within this call.

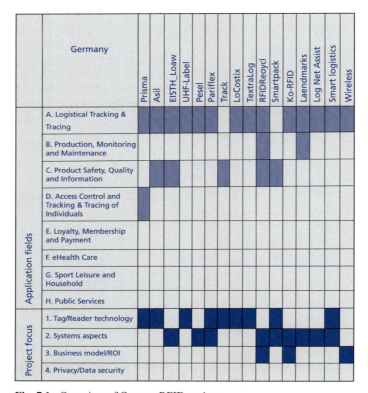

Fig. 7.1 Overview of German RFID projects

7.2 Analysis of National Programmes

Other Programmes: Two other calls within the "Mikrosystemtechnik" programme deal with technologies and applications that may have some significance for RFID from the technology and applications point of view:

- "Smart Technical Textiles"
- "Energieautarke Mikrosysteme" (energy self sufficient micro systems)

We could not identify specific RFID-related projects in those calls.

BMWi Programmes

Within the programme line "Next Generation Media", with a total call volume of 80 Mio. € (funded volume: 40 Mio. €), one of the four thematic clusters focuses on RFID. The theme cluster "Logistics networks – everything on track" supports four projects.

In the introduction to the programme launch, the BMWI pays special attention to the issue of involving SMEs into the rollout of the technology across all kinds of logistics applications and the bottlenecks encountered:

- Open technical issues (e.g. harsh environments/metals/liquid)
- Software/middleware architectures
- ROI/business case
- Security issues
- Standardisation

Four projects with RFID-related content weree identified. In addition, the BMWI supports the: "Netzwerk elektronischer Geschäftsverkehr" (electronic commerce competence network) initiative, which offers RFID-related support (information and consultancy) to SMEs planning to introduce RFID in their business processes.

Deutsche Forschungsgemeinschaft

The DFG funds seven fields of research. For all fields there is 363 Mio. € available. One of the special research areas with RFID significance is "Selbststeuerung logistischer Prozese" (self steering logistic processes). In this programme, which consists of a total of 14 projects, RFID is addressed as one of the main enabling technologies for the ICT aspects of this research area.

Fraunhofer Gesellschaft

In recent years, no projects with a focus on RFID were funded; the last project dates back to 1999: The project "assist" (personal shopping assistant).

Regional Programmes

From several regional programmes in Germany, the Stiftung Industrieforschung supports RFID-related projects. For 2007 a programme has been formulated around topics such as:

- New concepts for high reliability applications of RFID
- Development of methods to validate the cost/benefit trade-off of RFID applications.

These programmes are mainly aimed at small und medium-sized companies considering RFID. The programme focuses on aspects of the business case for SMEs and ROI aspects of this novel technology.

Also the Bayrische Forschungsstiftung (Bavarian Research Foundation) supports RFID-related projects. An example is the project CMOS-RFID-S, executed by the Institute of Electronics Engineering of the University of Erlangen. This project deals with passive, locatable multi-standard CMOS RFIDs with sensor functionality for mass applications.

Conclusions

Germany exhibits a multitude of subsidy programme lines; most of which are based on a competitive call system. There seems to be a reasonable match between RFID-related technologies and applications, and the past and present thematic priorities of the largest funding bodies, the BMBF and BMWi.

On the other hand, this thematic call system has its drawbacks, for example for the EUREKA programmes (which rely on national funding): the "right" project at the wrong moment (no fitting call open) could jeopardise funding, and thus the consortium.

RFID is in the focus (or at least an important part) of a number of programme lines. Examples are the NextGenerationMedia and Microsystems programme lines of BMWI and BMBF respectively (Bovenschulte et al. 2007, BMBF 2007a, BMBF 2007b, BMBF 2004, BSI 2005). High political awareness of the potential impact of RFID and the economic benefits are supportive to RFID-related subsidy programmes. Furthermore, the strong position of research institutions and University institutes in the RFID field and their good contacts to funding agencies help to generate a sizable RFID project portfolio.

On the downside there is little coherence between the separate projects, to the extent of redundancy and overlaps between projects and there is no central repository for the experience and information gained especially when no academic institutions with a RFID-centric research programme are involved. This is well recognised in a number of recent policy papers on medium-term R&D policy (IPTS 2007, Van de Voort, Ligtvoet 2006, European Parliament STOA 2006, BMWi 2007) and also specifically for RFID; suggestions range from support for structural support for RFID-relevant technologies to support of regional and thematic clusters.

7.2.2 France

The following ministries and governmental agencies are involved in RFID-related R&D support programmes:

7.2 Analysis of National Programmes

- **OSEO Anvar** focuses on the support of SMEs, specifically innovation & technology programmes.
- **The Ministry of Industry (DGE)** harbours the "Enterprise Competitively Fund" of the Ministry of Finance, and finances industrial R&D programmes, especially in the context of EUREKA and its clusters.
- **ANR (National Research Agency)** supports fundamental and applied research, and technological transfer from research institutions to industry.
- **AII (Industrial Innovation Agency)** supports large cooperative projects of large enterprises, which surpass their usual R&D effort and aim at global markets, creating highly qualified employment and export revenues.

Pôles de Compétitivité

Pôles de Compétitivité (PdC) are specific regional forms of cooperation on various fields, which usually have a selection procedure for cooperative projects in their respective fields of operation. They may receive government subsidies from

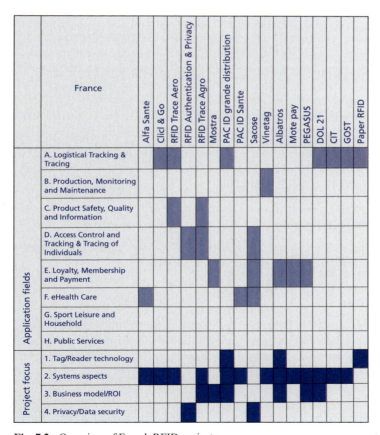

Fig. 7.2 Overview of French RFID projects

all sources mentioned above plus regional subsidy programmes and funds, depending on the character and scope of selected projects. This form of cooperation combines the advantages of regional clustering with thematic focuses in their programmes. There are concise guidelines from the French authorities and government concerning the endorsement of a PdC and financing is considered to be a long-term issue. Average subsidies amount to one third of total project costs. The subsidy funding comes from the above-named agencies (approx. 60%) with the remaining 40% from a specific government fund which bundles the budgets of all ministries concerned. Regional top up is foreseen. We found the following PdCs with explicit relations to RFID-thematic areas.

PdC SCS (Solutions Communicantes Securisées)

This PdC with its subordinate thematic groups contains the thematic areas, and associates industrial and scientific partners in order to cover a majority of RFID use cases.

- Identity
- Convergence
- Connectivity
- Mobility

By early 2007, 10 RFID-related projects were funded within this PdC.

PdC TES (Transactions electroniques Securisées)

TES is a regional PdC (located in Basse Normandie/Caen) and unites 16 large companies, 38 SMEs and three research institutions and concentrates on aspects of secure data transfer (technology, IT, services, etc.) that use case electronic payment. Three active RFID projects were found in the current programme.

PdC Logistique Seine-Normandie

This PdC focuses on logistic solutions, especially for containers and maritime transport. Again, three active RFID projects were identified.

PdC MINALOGIC

In this programme, located in Grenoble, the cluster combines micro-nano technology with software competence in order to foster innovation in intelligent miniaturised products and solutions for industry. Here, one active RFID project was found.

Conclusions

In summary, awareness on a national level for RFID is high (Roure et al. 2005). Both the structure and the size of the subsidy infrastructure in France are very beneficial for thematic clustering. Strong national support exists for industrial and application-oriented R&D, the amount of money available is considerable, and there is a strong focus on technology implementation; while still pre-competitive in nature, the programmes support activities quite close to final validation and market introduction.

7.2.3 UK

Department of Trade and Industry

In the UK the R&D priorities, and possibly linked subsidy lines are determined by the Technology Strategy Board, which defines relevant technology areas. These are delivered by the Department of Trade and Industry (DTI) and come along in two support lines:

- Collaborative Research & Development
- Knowledge Transfer Networks.

The total sum available over the period 2005–2008 is around 400 Mio. €. Calls within the Collaborative R&D framework are usually twice a year including an updated list of technological themes and addressing cooperation between companies and research institutions.

From a scan of the projects database, no specific RFID projects were found, with the exception of one project dealing with an RFID-related materials science project (printed electronics, high mobility polymer materials).

Conclusions

The UK subsidy scheme does not support larger companies outside cooperative (research-company cooperation) programmes. There is little evidence for targeted support for RFID-related topics, both in national or transnational programmes (such as Eureka).

On the other hand, the importance of support for "the last mile" in RFID introduction is clearly recognised. The Department of Trade and Industry supports the RFID Centre (www.rfidc.com), a public-private organisation focusing on advice and guidance in RFID matters such as business case development, technology testing, training and technical trials.

7.2.4 The Netherlands

Judging from the high number of official publications and papers (Ministerie van Economische Zaken 2006, RFID Platform Nederland 2006) from regional and national government agencies and ministries, RFID is one of the top innovation themes. Certainly, this is also enforced by the importance of the logistics and transport sector in the Netherlands.

The SenterNovem agency is the central institution for the coordination and the administration of the major part of national subsidy programmes. The agency reports to the Dutch Ministry of Finance. There is a multitude of programme lines, but none of which have a direct focus on the topic of RFID, i.e. there is no thematic programme for "RFID".

One of the major subsidy programmes supporting the cooperation of research institutions and companies is the so-called BSIK (Besluit subsidies investeringen kennisinfrastruktuur – Decision on subsidies for investments into competence infrastructure): within this programme line, two major projects were found which include some aspects of RFID technology and application (Smart Surroundings, Freeband).

A second subsidy line containing RFID-related topics focuses on the SME community; the subsidies are available in the form of tax reductions for R&D and innovation-related labour cost. Although a rather large number of SMEs have used this programme for mostly service related projects containing some aspects of RFID, the cash value of those subsidies is rather small (less than 1 % of the project cost).

A third "indirect" subsidy line for general RFID research comes centrally from ministerial side financing investigations and study reports, such as "RFID in healthcare", carried out in cooperation with medical institutions and market research companies.

Looking at plans for new programme lines, there seems to be a consensus that support for applications' investigations and prototyping could fill a gap on the way to a more rapid adoption of novel, RFID-based services. Interestingly enough, the Dutch government also recognises the leading role state procurement can play in supporting the adoption of RFID applications: several publications point at the importance of e-passport and e-transport card project for this technology, and suggest more initiatives in the future.

Additionally, the Ministry of Economic Affairs participates in industry and research institutions platforms, such as the RFID platform Nederland and eNederland, which aim at the dissemination of RFID application information and investigations of privacy and data security aspects.

Conclusions

In spite of the high interest in RFID and the potential importance of RFID for the key economic sector transport & logistics, there are surprisingly few subsidised projects in this area, considering the importance of the transport and logistics sector in the Netherlands. Presently, there seems to be awareness that there should be done more. Especially the "last mile" support, i.e. the support of real size testing and validation, should have more and bundled support.

7.2.5 Italy

From our search we conclude that in Italy national programmes in RFID technology and RFID projects in general are currently not subsidised. Nevertheless, Italian companies do participate in regional (EU-funded) and transnational (EUREKA) projects. One example is a recent project in Varese named REGINS-RFID, partly sponsored and subsidised by the European Community and several Italian Local Public Administrations (such as Lombardy Region, the Chamber of Commerce/Industry/Handicraft/Farming of Varese, the University of Castellanza-Varese and other local entities). REGINS-RFID aims to promote a new key technology for logistics to improve the transparency and quality of the logistics supply chain.

However, RFID technology is strongly promoted both in a private and a public environment. From a private point of view, the main sponsors are the University of Milan, School of Management of the Politecnico di Milano, and the Sapienza University of Rome. From a public point of view, the main sponsor for RFID-related research is the FUB (Fondazione Ugo Bordoni), a partly privately financed research centre affiliated to the Ministry of Communication.

Conclusion

National funding of R&D projects in general exists to a very small extent; there is no focus on direct RFID support. Indirectly, government-funded research institutions execute cooperative projects with the RFID industry.

7.2.6 Spain

In Spain most subsidy programmes are handled by one central agency, the CDTI (Centre for the Development of Industrial Technology). It is a public organisation under the Ministry of Industry, Commerce and Tourism, and it acts as intermediary between the ministry on one side and companies and R&D organisations on the other side. Financial support is granted by funds and loans of the government. Supported programmes come from all industrial sectors. While theoretically companies of all sizes are supported by these programmes, in practice there is a tendency to concentrate on SMEs. Programmes are usually divided into:

- Cooperative industrial research projects
- Technological development projects
- Technological innovation projects

There are no specific programme lines or themes that will be preferred in support so therefore there is no specific RFID-related programme line. Only three RFID-related projects in the CDTI portfolio (OMEDIS, NEPTUNO II, and RFID in traceability) could be identified; access to projects data is restricted.

Given the accentuated regional structure of Spain, there are also regional agencies working in industrial R&D subsidies. The Basque government, for example, sponsors the project KIROLTEK, which, amongst other things, tests the use of RFID for registration and timing in sporting events.

Conclusions

The basic subsidy structure in Spain supports companies and research institutions alike, including single company projects such as OMEDIS and NEPTUNO II. But according to our investigation there are no specific support programmes for RFID. One of the apparent weaknesses is the regional fragmentation of subsidy programme lines and the lack of focus on RFID.

7.2.7 Austria

National Programmes

In Austria, most national Research and Technology Development (RTD) subsidy programmes are financed, run and administered by two agencies:

- Fundamental Research Orientated: (FWF): Fonds zur Förderung der Wissenschaftlichen Forschung = Scientific Research Fund
- Applied and Industrial R&D and Transfer Programmes: (FFG): Forschungsförderungsgesellschaft mbH = Industrial R&D Promotion Fund

Almost all RFID-related activities fall within the scope of the FFG. The main FFG programme lines (and their link to RFID subsidies) are:

Basic Support Programme

This programme line focuses on applied R&D and supports R&D projects of individual companies and research institutions as well as cooperation. It addresses large companies and SMEs alike, although there is an ongoing discussion whether positive discrimination of SMEs should better not prevail.

Headquarter Strategy Programme

This programme type supports the transfer of core company R&D (and related company functions) to Austria, and addresses specifically internationally operating companies. Again, this programme type is not linked to any specific technologies.

Structure Programmes

This programme line supports the structural cooperation between research institutions and industry. Of the various programme types in this programme line, a very successful programme type in the past was the Competence Centre and Competence Network programme (This programme type is presently under revision which can mean a chance to bring more RFID content into the programme). The programme type supports consortia of Industry and Academia and research topics are principally open. At present RFID topics are almost non-existent in the currently operating industrial competence centres and networks. In the new follow-up programme (called COMET) RFID is a major topic in two of the proposals.

Thematic Programme Line

This programme line focuses on selected national thematic areas; selection and definition of these themes is frequently done by joint industry-academia panels under the guidance of the sponsoring ministry. Of the various programme types within this programme line, the programme FIT-IT ("Forschung, Innovation und Technologie für Informationstechnologien"/Research, Innovation and Technology for IT) has supported and still supports a number of projects of high relevance for RFID. Thematic areas relevant for RFID-related projects are:

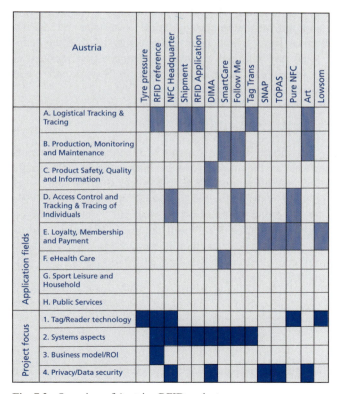

Fig. 7.3 Overview of Austrian RFID projects

- systems on a chip
- embedded systems
- trust in IT systems

In practice, existing cooperation between companies and academia in the RFID technology (e.g. NXP-TU Graz) have benefited from this programme line.

Specific ("Impulse") Programme Lines

Occasionally, ministries in Austria support special programmes focusing on specific economic categories (e.g. SMEs) in combination with a technology focus. One of the examples with full RFID relevance is the programme "Development and market introduction of RFID technology applications" run by the Ministry of Labour and Economy under the heading "Action programme for the digital economy/ICT". It addresses SMEs exclusively, and most likely is/was a one-off initiative.

Regional Programmes

As a mirror-image of the federal structure of Austria, the Austrian states have their own regional subsidy/R&D promotion organisations. There is a multitude of programmes, frequently not focusing on specific technologies, but supporting regional development and infrastructure programmes. Although no RFID-relevant regional programmes were found, their general way of working (supporting the creation of local and regional clusters and cooperation models) is quite successful in other areas (e.g. automotive cluster in Styria) and could serve as role model for future RFID-related structures.

Conclusions

The structure of subsidy programme lines in Austria is principally quite supportive to RFID-related projects as the major subsidy lines support both single companies and cooperation, are open towards technological themes, and cover the whole range from applied research to "close-to-market introduction". We only found a focus on RFID-centric programmes in one call, concentrating on SMEs. A larger and concerted action on RFID will require industry and industry/academia initiatives in defining projects within the programme lines, and cooperating with ministries and agencies in creating and defining specific programmes in support of the RFID value chain.

7.2.8 Finland

In Finland three funding schemes were found which were involved in RFID-related projects:

7.2 Analysis of National Programmes

Government Funding

The structure of R&D subsidies and the executing agencies in Finland have some similarity to the Austrian situation. The central agency dealing with subsidies for academic institutions and companies alike is TEKES, the Finnish funding agency for technology and innovation. TEKES's funds come from the national budget by the Ministry of Trade and Industry; their annual budget is around 500 Mio. €, and they fund close to 2000 projects annually (TEKES 2008). Funding is done by low interest rate loans and/or grants, depending on the nature of the proposed project and the stage of its innovation. Percentages for grants range from 50–100 % for research institutions, and 15–50 % for industries; additional loans are possible. The scope of TEKES project support covers both national and transnational as well as European programmes (e.g. EUREKA). The programme structure is twofold:

- An "open" system, similar to the Austrian base programme, with no thematic priority and open to submission throughout the year. About 60 % of the TEKES grants fit into this "open" funding line.

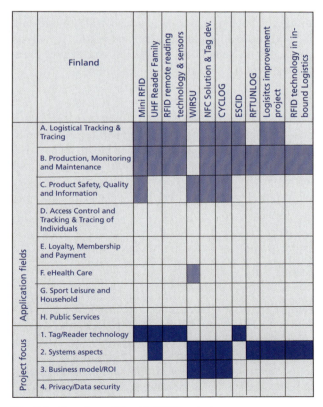

Fig. 7.4 Overview of Finnish RFID projects

- A more focused "technology programme" line, where, defined by regular reviews, the most promising technology areas are identified and project submissions fitting those areas and specific topics within these areas are solicited. Ideas for new technology programmes are based on initiatives by the "customer base" and on the focused areas regarding to the TEKES's strategy.

Partly publicly financed Contract Research Organisations

A typical example is VTT, which does cooperative R&D with Finnish and international companies, and receives about 20% of its funds from the government (via TEKES). In the project database we found 20 RFID-related projects, demonstrating that RFID is one of the larger topics in VTT's contract research portfolio.

Public-private Partnership Model: the RFID Lab

The recently opened RFID lab in Helsinki is an interesting example of cooperation between companies being active in the field of RFID (e.g. Nokia, UPM, and HP) and local governments and their technology agencies. The lab will provide consultancy services for system specifications, implementation and testing. It will run a demonstration and testing facility to support the validation and market introduction phase of RFID technology and solutions.

Conclusions

The Finnish subsidy model is very pragmatic, in principle open to all technologies and companies and research institutions alike. Technological focus areas are defined in close alignment with industry. Judging from the themes and topics, plus the presence of major players in RFID within these programmees, the Finnish subsidy model is rather supportive for RFID-related programmes and technologies.

7.2.9 New Member States

For our overview of new Member States, we selected the Czech Republic, Hungary and Poland. Our research showed that no national RFID R&D programmes exist in these Member States, however one Czech company participated in the EUREKA project and Hungary and Poland both participated in some FP6 programmes.

Although only those three countries were analysed, the lack of national programme focus on RFID and related technologies indicates a clear gap between the new Member States and the other countries selected in this survey.

7.3 Transnational Programmes with National Funding

7.3.1 NORDITE

One of the best examples of transnational cooperation between national authorities and programmes relevant to or focusing on RFID is a cooperation between the Finnish (TEKES), Swedish (Vinnova) and Norwegian (Research Council of Norway) funding authorities, called NORDITE.

NORDITE aims to support research institutes and universities in their effort to develop state of the art research in the fields of SW radio, wireless sensors, short range wireless networks and RFID.

In order to establish a link between the subsidised research institutions and future economic exploitation, the market potential of these projects has to be confirmed by at least two companies who have signed an agreement of participation. The programme runs from 2005–2010, the total subsidy volume is 16 Mio. €; up to now 38 companies and approx. 20 research organisations are participating. We found two examples of NORDITE-funded, RFID-related projects: printed RFID sensor solutions, and a project on integrated sensors (Intellisense RFID).

The fact that RFID projects are part of the transnational initiative is based on the fact that RFID technology is recognised as important for the core industries (manufacture, automotive, transport and logistics) of those Nordic countries.

The Nordic initiative is an interesting example for transnational cooperation, initiated by national authorities and focussing on areas of common industrial and business interest. Given the fact that RFID as technology and application will need concerted efforts across national boundaries, this initiative can serve as a reference model for implementation of a regional R&D policy beyond pure national priorities and interests.

Table 7.2 Overview of NORDITE programme

Number of national agencies involved	Number of dedicated RFID program lines we found	Number of RFID-related projects we analysed	Project volume of RFID-related R&D support
3	1	2	Not available

7.3.2 EUREKA

The Eureka programme in its two forms, bottom-up and cluster projects, addresses transnational cooperation and relies on national funding of individual participants. It will be most successful in countries which have either earmarked budgets for participation in EUREKA, or a subsidy programme line that neither stipulates specific technology priorities, nor operates a competitive call system on a national level.

EUREKA Clusters

EUREKA clusters closest to RFID-related technologies and applications are:

- CELTIC
- EURIPIDES (a merger of the cluster programmes EURIMUS and PIDEA)

CELTIC deals with integrated telecommunications systems, wherein RFID could play a major role in certain aspects. In contrast, EURIPDES focuses on smart systems integration, in which all packaging and systems integration aspects of RFID applications come into play. Scanning through available project files, no RFID-related projects were found in CELTIC calls until now. Within EURIPIDES and its predecessors (with their focus on smart systems integration), two packaging projects were found. They relate to specific, multi-layer and wafer level packaging designs for smart cards.

When searching for bottom-up and umbrella projects, 11 projects with RFID content were found, for the period between 1998 and 2006 in the EUREKA programme files.

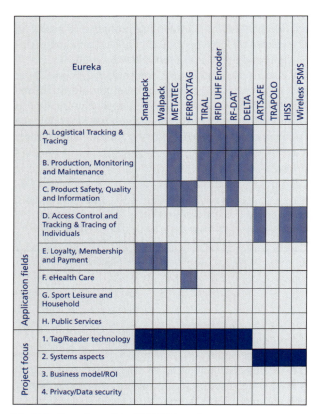

Fig. 7.5 Overview of the EUREKA project

Table 7.3 Overview of the EURKEA project

Number of national R&D support agencies involved	Number of dedicated RFID program lines we found	Number of RFID-related projects we analysed	Project volume of projects we analysed
	none	12	

Conclusions

Generally speaking EUREKA is a bottom-up, low overhead programme and as such quite successful in stimulating trans-national R&D, although there is no specific RFID-centric cluster within EUREKA, the existing cluster programmes EURIPDES and MEDEA could function as a home base for the microelectronics and systems integration aspects of RFID.

The main drawback is the reliance on national funds for financing, and the necessity of alignment of sometimes very differing national funding rules to start a project. One step to overcome this problem is the EUROSTARS programme co- financed by the EU and requiring earmarked funds from the participating countries. EUROSTARS could, in the future, support RFID projects with SMEs as main participants.

7.4 Transnational Programmes with Joint National and EU Funding

Interreg III

Some programmes which are funded by the EU to stimulate and support regional cooperation and development may create infrastructure for specific regional technology focus. For example, Regins, (a regional framework operation project within the Interreg IIIc programme) focuses on cluster management and knowledge transfer to regional SMEs. Its thematic priorities are automotive, logistics and biotechnology.

Within this newly created infrastructure, the subproject REGINS-RFID has been defined, with participants from Germany, Hungary, Austria and Italy, focusing on the promotion of RFID, an analysis of the current status of RFID, a development of guidelines for implementation, and especially on the needs of SMEs.

7.5 European Programmes

The Framework Programme

At the EU RFID Conference "Heading for the Future" (Brussels, 16 October 2006), the Directorate General for Information Society and Media published a concise overview of the RFID portfolio of European research and a shortlist of RFID-related projects from FP4 to FP6. (DG INFSO 2006)

Two graphs taken from the overview (DG INFSO 2006) give an interesting analysis of the distribution of those projects among research domains and among RFID application areas:

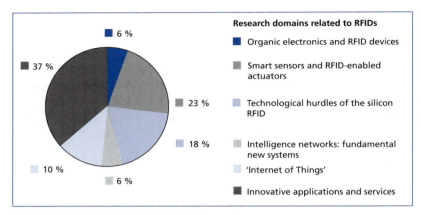

Fig. 7.6 Research domains related to RFID

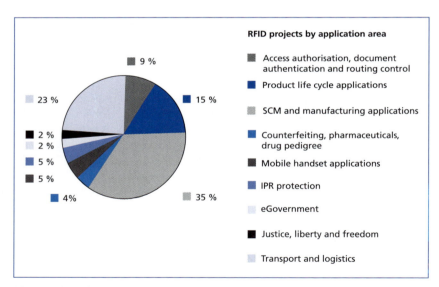

Fig. 7.7 Overview of application areas in European programmes

An interesting first conclusion is the relatively large share of innovative applications and services (37% of the research domains) and the large proportion of the SCM, manufacturing and logistics applications (close to 60% of the application areas). On the other hand, the total cumulative project budget of all RFID-related

7.5 European Programmes

Table 7.4 FP5/FP6 programmes

Programs we analysed	Number of dedicated RFID program lines we found	Number of RFID-related projects we analysed	Total volume of RFID-related R&D support (Mio €)
FP5+6	none	20	168

projects (312 Mio. €) and EU subsidy level (153 Mio. €) is a relatively small share of the total FP6/FP5 budget.

In view of the present importance attributed to RFID as a European core technology, there is surprisingly little coherence between the projects and its consortia. No past initiatives were found which formed R&D competence bases involving companies in the specific field. In the run up for FP7, this issue has clearly been recognised: the creation of ETPs (European Technology Platform), bringing together the key stakeholders in a specific domain, allows the bottom-up definition of a Strategic Research Agenda (SRA), which in turn provides input for the FP7

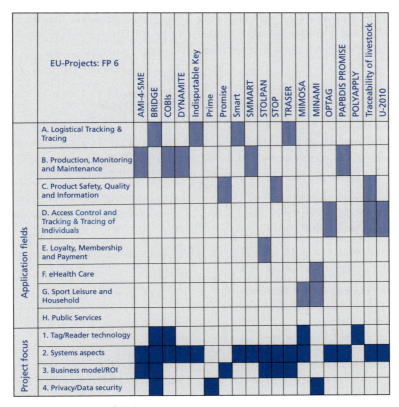

Fig. 7.8 Overview of FP6 programmes

programme. Implementation of such an SRA can be done, in principle, through collaborative R&D in FP7 or other resources. RFID is to a certain extent covered in the SRA of the European Technology Platform EPOSS.

As a short-term measure, a number of current FP6 RFID projects are currently organised into a strategic project cluster (CERP: Cluster of European RFID projects), with the explicit aim to coordinate research activities and establish synergies between the projects. Although this initiative seems to have been a temporary ad-hoc measure to answer criticism of RFID-related framework projects (especially concerning a lack of coherence between project goals and their consortia), it certainly is a positive development.

Another more generic criticism concerns the relatively high administrative overhead for the framework projects, combined with a rather low hit rate (probability of being accepted). Whether this has been addressed in FP7 remains to be seen.

Conclusions

Compared to the earlier framework programmes, FP7 clearly gives more weight to the view and priorities of industry. Assuming that the numerous discussions on RFID-related topics will translate into focused themes and topics, RFID and its technology base will play a more prominent role in the project portfolio. The question how to effectively cluster and make accessible the RFID-related competence generated in those programmes still remains an open issue.

7.6 R&D Programmes & the RFID Reference Model

As previously explained, the RFID Reference Model has been used for this analysis to allow project themes to be highlighted, so that areas where little research is being carried out can be identified. In this section we try to answer the following questions:

- Is there a clear trend in the application focus of supported RFID R&D projects?
- To what extent do the national and European subsidy programmes address the bottlenecks identified in Chap. 6?

For this analysis, we map the project to the RFID Reference Model with an extension relating to the project focus:

- focus on the tag and reader technology
- focus on systems and application aspects
- focus on business case
- focus on data security and privacy issues

As has been previously stated, the access to specific project information on national level was rather difficult, so our database as given in the project mappings on the extended RFID Reference Model is based purely on the information available. However, we can distinguish some indicative trends.

As to the application focus, the majority of the projects fall into the RFID reference model categories 'logistics, tracking and tracing', 'production, monitoring and maintenance', and 'product safety, quality and information'. This is not surprising and correlates well with the relatively large number of practical use cases in today's available RFID applications.

The project mapping on our bottleneck criteria allows to draw the following conclusions:

- A tag/reader technology focus is well represented in national subsidy projects and EUREKA bottom-up projects. Looking more closely at the project descriptions (to the extent available), the focus is frequently on added functionality (sensor integration) and achieving lower costs via novel technologies (e.g. printing).
- The system aspects' focus matches the tag/reader focus in national programmes and dominates the EU projects. Local/geographical clusters (e.g. France) seem to favour a system approach.
- Generally speaking, the business case focus is underrepresented. In most projects in our survey it is treated just as one of many topics. There are very few projects with a sole focus on the question of how to achieve a conclusive business case model.
- Privacy and data security: only a few projects, and mostly with a focus on "privacy by design" or data security in complex networks. Only two FP6 projects focused exclusively on privacy (PRIME and MINAMI).

From the analysis, it has been determined that the last two points, concerning establishing business cases and privacy and security, will need the most attention in the near future, as it has been found that these are two major bottlenecks which prevent the development of RFID.

7.7 Conclusions of RFID R&D Funding Programme Assessment

7.7.1 Thematic Focus of Funded Programmes

Although it is sometimes difficult to obtain detailed information on the content of subsidy programmes (especially on a national level, due to confidentiality issues in some countries), we found some general trends as to where the different programme lines focus on.

National programme lines and especially those not based on thematic calls, but on bottom-up project proposal systems, tend to support programmes closer to market introduction, dealing with topics like robustness, reliability and cost. This trend is even stronger when the project consortia are located in a regional cluster and represent companies along the value chain, as in a number of French PdC (Pôles de Compétitivité) projects.

Thematic calls usually focus on novel systems' aspects and device innovations required for their implementation.

The past EU framework programmes show a similar trend: from a technological point of view, the majority of RFID-related projects concentrate on system innovation and to a lesser extent on device technology.

For this analysis of the EU project portfolio, a time period from early 2000 to early 2007 was used, and it was discovered that earlier projects tend to have a higher technological content, while more recent projects concentrate more on novel system solutions and their application to advanced business processes, including studies and development of business cases.

Generally speaking, few programmes explicitly focus on privacy issues (only two projects dealing with privacy were found), while the issue of data security is a topic in several others. This is all the more surprising, as this does not reflect the current discussions and to some degree the controversy around RFID and privacy concerns.

When looking at different national R&D programmes there are only a few dedicated to RFID explicitly. Two examples (which are described in more detail in this report under their respective titles) are the Impulse programme in Austria that orients towards SMEs, and the call "Microsystems Technology for Smart Label Applications in Logistics" within the Micro Systems Technology programme of the German Ministry of Education and Research.

In general, programme dedication to RFID and/or to topics related to the use of this technology seem to reflect the importance a national R&D policy gives to this technology and the role the respective national industry wants to play in the European economy.

7.7.2 Funding Structures

The level and type of support varies between the countries we identified as key countries for RFID in our analysis:

Germany and France are in the lead concerning direct subsidies to companies, and subsidise almost exclusively cooperative programmes. The UK and Italy do not (or rarely) subsidise industrial company R&D directly; their support to RFID runs with the aid of programmes in publicly financed universities and research institutions, which in turn cooperate with companies. The same seems to be true for the Netherlands; although in the past there were specific arrangements with large Dutch companies which most likely also covered RFID-related R&D.

Direct single company R&D projects receive support in Austria and Finland, but due to privacy agreements only very restricted access to project information was available.

The status of RFID-related R&D in the new European Member States is difficult to analyse. There is no evidence for national programme lines in the Czech Republic, Slovakia and Hungary. Some companies, however, from these countries participate in EUREKA and EU projects with an RFID focus.

7.7 Conclusions of RFID R&D Funding Programme Assessment

National R&D policy will focus on improving the position of national players. However, the success of RFID in many applications e.g. in the field of supply chain management and logistics needs a transnational/European approach. We found one example of regional cooperation of national funding bodies (NORDITE) supporting amongst others, transnational RFID R&D, an example which is worth further studying.

Apparently due to the restrictions of the existing state aid rules (in their original more restrictive formulation) most programmes and projects do not go beyond technology demonstrators and prototypes; in practice, a considerable development effort still has to be made to come to a functioning and economically feasible system; and in our judgment this is still 'pre-competitive' and highly innovative. In two countries (Netherlands and UK) initial discussions and ideas were found on side of government agencies about how to better close the gap between development and validation. In Finland a partly publicly funded agency was founded to close this gap.

Looking at the French example, supporting regional clusters (and the recent ideas to make those clusters transnational) seems to be a formula for success to promote cooperation and networking along the value chain of innovation topics. This also applies to the RFID technology.

Within those clusters RFID can be seen as one of the thematic areas, and in the example of the two PdCs in France, RFID projects labelled by the cluster are eligible for national funding. The real strong point of this system is the availability of earmarked funds for cluster projects (and thus for RFID-related topics, if those are focal themes in the clusters themselves).

Many of the RFID projects we found are cooperation between large companies and SMEs and many of them also with participation of public research institutions (one exception is a programme exclusively reserved for SMEs). On the other hand, there is an obvious tendency in many countries to favour SMEs in public funding schemes. From our point of view, part of the success for SMEs lies in cooperation with larger companies e.g. by improving market access etc. Therefore the public programmes should not discourage the participation of large companies in subsidised R&D projects. As there are many successful small and medium-sized companies in the RFID field, this applies all the more for this technological area.

The issue of creating a central RFID competence repository, either national or European, has not been addressed in the past, but is discussed in various ways by national authorities and industry alike. In most scenarios research institutions will play an important part, but the model of public-private partnerships for the development and validation of generic use cases will need further attention.

Concerning RFID within the EU framework programmes, there is little coherence between projects and project consortia. There even seems to be a certain overlap between project topics and goals. Given the Europe-wide focus on RFID, some kind of clustering would be beneficial. The strategic cluster CERP is a short- term means to address this point. One of the more strategic actions to support RFID would be the support for ETPs with a clear RFID technology and application focus.

7.8 Recommendation for a Future European R&D Policy

Regarding recommendations in the field of RFID it is important to be aware of the phase the technology and its applications are in: in major areas it has left the realm of basic and applied research and is on its way into the market.

As indicated in the introduction to this book, RFID and related technologies are already well in the innovation phase. When referring to the technology adoption model of Rogers (1962) and Schumpeter's view on research and innovation (1939), one can see that RFID technology is in the innovation stage, which means that R&D policy support for RFID should create the right environment and boundary conditions to make it a successful innovation (i.e. allowing to create value and allowing its proponents to make money with it).

On the other hand, there is a consensus, and not only within the academic research environment, that we have barely scratched the surface of what is possible with this technology, that there is an incredibly broad range of potential applications, which also means that there are great technological challenges and open research issues to keep applied research occupied for the coming years.

From a vantage point, this situation can be described best by what has been coined the open innovation environment, where research, application development, prototyping and market introduction not only co-exist, but exhibit intensive interrelations promoting quick transition from research to application, rapid and effective testing of novel ideas, and backflow of market and application experience to research. Thus an effective R&D policy in support of RFID should create the right environment for such an open innovation scenario.

Concerning technological themes and topics for R&D programmes there are two areas to be considered. First, partly due to the rapid growth in specific application areas of RFID, there is an apparent lack of consistent methodologies, support tools and simulation and compliance test tools and methods. Application-specific developments lack a solid scientific basis; repairing this deficiency should be part of the thematic focus of R&D programmes, as this is a major stumbling block for rapid and cost-effective market introduction.

Second, RFID technology requires a solid understanding of many disciplines, from information technology to signal processing, semiconductor technology, materials technology, RF antennas, privacy and security aspects etc. Programmes focusing on advanced aspects of those technologies and their cross disciplinary aspects will be the basis for the successful development of advanced and novel RFID applications in the future.

Finally, a European RFID R&D policy can help to increase public acceptance of this technology and its applications by linking support for RFID (both in research and application development) to its contribution to solutions for major societal problems, e.g. in the fields of healthcare, security, or environmental management, and by underscoring the importance of privacy by sponsoring additional specific (and not exclusively technology-oriented) programmes.

7.8 Recommendation for a Future European R&D Policy

Recommendation: Create and support an open innovation environment

An open innovation environment is characterised by:

- Close cooperation and frequently co-location of research institutions, large companies and SMEs.
- Rapid flow of ideas up- and downstream the R&D chain: Successful innovation requires interaction and iteration.
- Concurrent activities in research, development, prototyping and market testing activities.

In such a setting it is rather difficult and very counterproductive to differentiate between programmes according to the state aid classification as to maximum permissible aid. Recent developments in the state aid issue show clear improvement over the older, more restrictive rules. Examples are: the relaxation of rules concerning the classification of academic participation in joint projects as state aid, and the inclusion of prototyping in the activities eligible for funding.

To make such a cooperation successful, it needs critical mass of academic and company R&D personnel, a trusted environment as to IPR issues, and stable rules of cooperation, including funding and financing for medium- until long-term. We therefore think that all measures that enable and support industry and academia cooperation are of paramount importance to create this open innovation environment.

Judging from experience of companies in regional cluster programmes, geographical proximity of partners is beneficial, as successful regional and cross-border innovation clusters show. Generally speaking, clusters prove to be a very effective innovation environment. One of the reasons certainly lies in the fact that clusters are formed by bottom-up initiatives of research institutions and regional companies and/or companies with the same or complementary business interests, common technology and market vision. This is often more effective than top-down definitions in thematic calls, which stimulate more or less ad-hoc interest groups for a limited time period.

Therefore, we would recommend that the EU supports the creation of regional and transnational clusters to enable an open innovation environment.

- By means of not discriminating specific partners, e.g. large companies by excluding them up front from the participation in certain funded programmes. The cooperation between large companies with their market knowledge and market access and SMEs with their flexibility and focused innovativeness is a formula for success.
- By defining programmes dealing with the creation and operation of a European research infrastructure in a way to solicit and increase the participation of industry. This was not the case until now, with the research infrastructures being part of the framework programmes; integrating industrial private research infrastructure in such an open innovation setting certainly will be beneficial for both industrial and academic research.

- By means of allowing/encouraging national support for regional cluster formation, e.g. by national support for infrastructural measures such as jointly owned and operated application testing facilities.

> **Recommendation:** Support Europe-wide RFID deployment

Our analysis has shown that the particular focus, emphasis, or support concerning RFID technology and application differ widely across Europe. New Member States and CEE (Central and Easter Europe) countries seem to lag behind, especially in pre-competitive applied R&D programmes. On the other hand, in order to reap the benefits of this technology, a Europe-wide balanced introduction is needed in many areas, e.g. public transport, international logistics and supply chain management, public security, or healthcare etc. The current Framework Programme (FP7) will most likely not be able to close this gap on its own, although that companies and research institutions from CEE countries participate in framework projects.

Europe should check which instruments in its R&D and innovation policy portfolio are suited best to stimulate regional participation in this development. One of the likely candidates amongst the EU programmes for that endeavour is the Competitiveness and Innovation Framework programme (CIP) (CIP 2007), more specifically the ICT Policy Support (ICT PSP) part of the programme.

Therefore, we would recommend that the EU investigates the application of the CIP/ICT PSP toolset to promote the uptake of RFID technology and applications in areas and services of public interest within the CEE region and new member state countries. By these means, competence and awareness for this technology should increase in local companies and institutions. This will also help to bridge the culture and language gap with local CEE SMEs and support their integration and participations in transnational programmes.

> **Recommendation:** Balanced support for R&D themes and topics

In this recommendation, we want to address two distinct points. One refers to the content i.e. what topics should be supported (the "what") and the second point refers to the way how support should be given to be most efficient (the "how").

What should be supported?

Recently a number of studies and papers have appeared presenting, amongst other points, an inventory of R&D topics that Europe's industry, academia and government authorities should focus on. During the course of this project, and in parallel to our analysis of the European RFID subsidy programme scene, a list of potential R&D topics was generated, starting from a technological bottleneck analysis also within this research project (Wiebking et al. 2008).

We do not want to discuss detailed R&D items in the context of this recommendation, but we will point out that there is a general agreement on R&D theme clusters, which will need further attention. It should be stressed that although the R&D funding agencies currently favour mid- to long-term research programmes, short-term research projects must also be supported in order to efficiently further RFID development. Please be aware that the term "short-term" does not imply that the topic is of passing interest and importance and has little relevance in the long term. By saying "short-term" we want to indicate that the issue is of high urgency, work is needed now to clear immediate make-or- break bottlenecks. We will illustrate our recommendations by giving some typical examples:

Short-term:

There are several recommendations which fall into the tag/reader technology cluster. Initially, it is recommended that there should be focus on solving shortcomings of present (UHF) implementations, such as improving operational reliability under difficult environmental conditions (heat, metal or liquid environment). Focus on improving readout ranges is also recommended besides providing solutions for false positive readouts and the multi-reader environment issue etc.

And all of this should be carried out in a predictive, not case-by-case optimisation/trial and error approach, which implies more research in RF and antenna design, predictive modelling and emulation studies and similar issues, which might receive less attention in comparison to long-term programme points.

Short- to Medium-term:

In the tag/reader technology field, it is recommended to focus on low cost via IC design breakthroughs in the short-term, mass production technologies for antenna/label manufacture, integration of the tag function into the packaging, and (medium-term) novel (non-silicon) technologies for the integral tag function. Attention must also be paid to developing low power consumption for tags to further improve readout range and allow passive tags where the energy for added functionality is taken from the electromagnetic field of the reader. Added functionality, such as integration of sensors (temperature, pressure, etc.) or addition of bi-stable displays is also an area which should be researched.

Further research into system design is also required, such as developing system integration possibilities by standardising interfaces to system middleware and the standardisation of application layers. System and software architectures which will allow the transformation of data collected from smart tagged objects to business-relevant data should also be developed.

The question about already existing business cases and positive ROI for all partners in the value chain as an important motivation to adopt the technology appears in most studies. The lack of validated experience and accessible data, especially for SMEs is addressed frequently. Typical suggestions relate to specific SME-focused programmes in "Centres of Excellence", similar to the public-private partnership initiative in Finland, to support them by developing their specific use cases and R&D on simple, low cost, open source systems to lower the investment and start-up cost barrier for SMEs.

Another Research field that should be tackled on shorter notice is the area of privacy and security. Further research into privacy by design such as data encryption, access rights management and reliable deactivation methods is required. As said before, privacy is not a technology – only a topic, but adequate technology is a necessary prerequisite.

Research also needs to be directed towards studying the impact of RFID implementation on European citizens, and guidelines should be developed for the industry. Information and communication strategies should also be developed to facilitate a positive and proactive approach towards RFID.

Long-term:

RFID as a whole is a very complex system technology which requires inputs from many technology fields, e.g. information technology, electrical engineering, signal processing and communications, data security etc. Therefore it will benefit from general advances in those areas. The common focus of applied research in this area has to come from technology roadmaps, e.g. as proposed and demonstrated by Wiebking et al. (2008), and from proposals for new applications with their specific requirements profile. There is a multitude of system- and application-related topics which will have to be addressed in this context. This requires a holistic approach. Bilateral/multilateral projects of limited duration, while essential to progress, selected chapters of technology and applications, cannot be the only means.

How should this be achieved?

Generally speaking, the long-term vision part reasonably matches the focus on advanced technologies and systems of the current FP7 programme. One of the drawbacks of FP7 projects is certainly the one-time nature and the lack of cohesion between projects around similar topics. Temporary clusters such as the Cluster of European RFID Projects (CERP) are a step in the right direction, but a more permanent setting is desirable. It is therefore recommended to create a permanent project cluster covering all European programme lines as this would be a step in the right direction. Additionally, we would recommend using the model of European Technology Platforms (ETPs) and their Strategic Research Agenda to come to a structured and lasting focus on RFID-related technologies and their applications. Two existing ETPs explicitly address at least aspects of RFID:

- EPoSS from the smart systems integration point of view
- ARTEMIS in its Strategic Research Agenda on seamless connectivity and middleware

These ETPs can provide a solid basis to turn long-term strategic RFID topics into joint R&D programmes, addressing available national and European funding possibilities. On top of the thematic match, ARTEMIS also addresses the topic of innovation ecosystems as one of the core enablers for its SRA, using existing regional clusters to build Centres of Excellence. In addition to this, the declared focus on the involvement of SMEs comes very close to address core points of the "how to" wish list of RFID.

Looking at the short- to medium-term technology and system issues, some of these topics may appear to be straightforward product design and implementation issues, and thus not really eligible for sustained funding support or at the lowest possible level specified in the new state aid rules. In reality, major advances in those fields are urgently needed, thus requiring close cooperation of companies and research institutions, to shorten the presently lengthy and extremely costly "trial and error" design process and give quick and reliable answers as to the validity of business cases in new implementations.

A part of the answer to the privacy concerns will come from studies and projects in the "privacy by design" area which is a very important topic in the short-term research portfolio. Therefore, we would recommend that future R&D efforts shall target also these short to medium-term topics, in order to make them eligible for public support i.e. national or by the EU.

The problem of not having a central competence repository accessible for all players in the value chain needs a specific approach: We recommend that national or European authorities will (co-)finance application labs in form of public-private partnerships in order to support testing, validation and certification of technologies and concepts, and to generate relevant information relating to business case issues.

In some countries organisations, covering parts of the scope outlined above, either have been created or are in the planning stage; supporting these to meet the wider scope and giving them an "official" status, could give a head start in this direction. Such institutions can have an additional benefit related to public acceptance of RFID technologies in daily life by giving conclusive answers to privacy concerns. These institutions can provide compliance checks of technical and systems solutions against agreed standards and thus act as an impartial and independent authority.

Recommendation: Support "go to market" of novel applications

European initiatives in supporting R&D and innovation mostly deal with the "supply side" of innovation i.e. creating and supporting R&D infrastructures and generating technological competence but rarely deal with the "demand side" for those innovations. If, for various reasons, there is little demand for new products and processes, there is little interest in investing in the input side.

In this recommendation we intentionally take a broader approach to the issue of "R&D policy": in our opinion, any measure that stimulates industry's R&D spending on technologies and applications of economic and societal relevance, be it cofinancing of projects, establishment of a responsive research infrastructure, or creating a favourable application environment, is an integral part of policy in support of R&D. This approach is in line with the core of the recommendations of the so-called Aho Report (Creating an innovative Europe) and the so-called Wilkinson Report (Public Procurement for Research and Innovation), stating amongst other topics the need for Europe to provide an innovation-friendly market for its businesses, with actions on regulation, standards and public "procurement for innovation" (Aho 2006, Wilkinson 2005).

RFID may serve as a typical example: As mentioned above, new RFID applications with their specific requirements may spur research and technological advances in various areas of general importance to Europe's innovation drive. On the other hand, given the lengthy process from idea to market introduction, and lack of incentives due to highly fragmented national markets and their regulations, private companies may be reluctant to take the initiative and the cost and risk burden.

A typical example in the context of RFID is tagging of pharmaceuticals and medications in hospitals. On the upside, this would prevent counterfeiting and reduce the risk of wrong dosage and medication. However, presently the national diversity in regulations in that area makes widespread adoption of this RFID application a major business risk for all players involved. As a side effect for the European industry, not to invest into this novel application would mean losing out in technological areas related to such applications.

Here, European initiatives could help; clear recommendations on European and national level, comparable to the US FDA recommendation ("combating counterfeiting drugs") will support decisions of companies to invest in this area. Opening up public procurement for new technologies and thus not only looking at lowest price or no risk is another way of promoting novel RFID-related technologies. The tremendous success of passport applications, which is basically a windfall profit from enhanced security requirements worldwide, is a perfect example of public initiatives driving technology development.

Therefore, we would recommend that Europe takes the initiative to promote RFID applications in areas of societal importance e.g. drug authentication, efficient logistics and transport, automotive industry, or transnational use of eGovernment techniques by supporting the application of RFID-based solutions. State procurement should open up for novel, potentially risky solutions to foster technology development and application areas where RFID technology can provide the answers. In a recent communication to the European Parliament the European Commission has already outlined such an approach, by discussing the ground rules for the so-called "pre-commercial procurement", the procurement of R&D services on a non-exclusive and competitive basis to shorten the time-to-market of new and promising technologies (COM(2007) 799 final). This policy should be put into practice in general and be applied to RFID specifically.

Chapter 8
Conclusion: The Next Steps for Europe

RFID is a major opportunity for Europe. It provides societal benefits in many applications in areas such as health care, food safety and environmental monitoring. Users from the private and the public sector may use this technology to improve their efficiency and to provide new products and services based on RFID. For technology suppliers and system integrators, RFID offers the opportunity to participate in a new and growing high-tech segment of the ICT market. However, a closer inspection shows that for most applications RFID is still in its initial phase of technology adoption. In order to enter the second phase of broad technology adoption, several challenges must be met.

This chapter summarises the results of this book – and of the European coordination action project CE RFID – on these challenges. The recommendations for further action reflect the point of view of two important stakeholder groups: RFID users and technology suppliers. They aim to further extend the European position as innovative key player in the field of RFID technology.

We will first present the findings according to the fields of activities discussed in this book: standards, implementation and application guidelines, regulative framework, technology research needs and R&D environment. Secondly, the recommendations are regrouped with respect to the various stakeholders groups which are most often concerned with more than one field.

8.1 The Fields of Activities

Standards

Large ICT infrastructures depend heavily on standards, be it de-facto or de-jure standards. This holds especially true for RFID, since firstly a great number of areas from the air interfaces to data formats are affected, and secondly the building of efficient and economically priced networks depends on common technical conventions. It is important to, wherever possible, create broadly and internationally accepted standards which can be used for as many applications as possible. These should be open standards free from over- or under-standardisation. Basic standards which regulate frequency allocation, air interfaces and tag data specifications should be in the main focus. Interoperability of standards should always be kept in mind. Some specific applications require additional standardisation. In cases where generic standards would lead to suboptimal results, application-specific standards should be developed ensuring interoperability with the existing framework of standards.

Another important topic is to ensure an appropriate radio spectrum framework. Europe is at a disadvantage in comparison to the United States or Asia, because European radio spectrum ranges are smaller, and there is a lack of harmonisation. Cooperative and harmonised steps are urgently required to maintain Europe's position as an international competitive player.

Implementation and Application Guidelines

Our analysis has shown that due to a broad scope and a lack of sufficient information, existing guidelines are often of limited practical use for RFID implementation. Commonly applicable guidelines in terms of "one-size-fits-all" for the technology as a whole are too broad to meet the informational needs of different users. That means that besides the aspect of being application-specific, guidelines also need to be written with the addressee in mind, knowing that different addressees demand different information, for example an IT Department needs technical information, in comparison Human Resources requires information on how to inform employees. When drafting guidelines it has to be clear who the guidelines are written for and what application they are aimed at, as to integrate all informational needs of the target groups.

To help writers of specific guidelines, an addressee-specific checklist has been elaborated. This RFID Implementation Checklist attempts to list all relevant points that have to be considered by different addressees within the user company. Authors of new RFID guidelines can use this checklist as a guide to write their individual application- and addressee-specific guidelines. The advantages of checklists are that they allow a balance between sufficient guidance on the one hand and the necessary flexibility for specific use cases on the other. To ensure that the list can be a sufficient basis, further discussion and regular updates are necessary. This could be realised by RFID-specific business organisations.

Regulative Framework

Besides the technological and standardisation aspects, the regulative framework plays a crucial role for the deployment of RFID. This covers a number of topics, including data protection, spectrum management, environmental protection and patents.

Premature legislation could hamper otherwise promising RFID development. The existing legal framework regarding data protection suffices for RFID technology applications: as any other technology, RFID shall comply with the Data Protection Directive and related legislation, as well as with Member States' national data protection laws. The privacy impact of RFID should be evaluated application by application. Comprehensive public information and a continuous dialogue between the European Commission and the Member States is encouraged in order to monitor the application of RFID with regards to privacy, but also in other areas such as health and environmental aspects. Self-regulation and guidelines should be welcome. Premature legislation could hamper otherwise promising RFID development.

Additionally, the European Union should keep up its commitment to harmonise spectrum management and allocation across the Member States. Additional funds for technological research on these topics shall be welcome.

The European Commission and other stakeholders have established a coherent position on how to treat RFID under WEEE (Waste Electrical and Electronic Equipment Directive) and RoHS (Restriction of Hazardous Substances Directive): although RFID principally falls within the field of both directives, continuous monitoring of technology development is advisable.

Furthermore, the European Commission should continue political efforts to come closer to a Community Patent, encourage European participation in patent pools and standardisation initiatives and bodies and establish regular consultations with US and Asian representatives, creating a lively and innovative IT sector by allowing for different approaches in how to manage IPR, from open source to patent pools to other industry agreements.

Lastly, the European Commission should keep a close contact with all relevant stakeholders from industry, government, civil society and academia alike, as for instance represented in the RFID Expert Group and the High Level Group on Internet Governance in order to foster an open, transparent multi-stakeholder dialogue on RFID governance.

Technology Research Needs

Although RFID is a mature technology for many applications fields, there is still a strong need for further development. New requirements from new application fields often reach beyond today's technical limits. Current application roadmaps reveal many opportunities for evolutionary improvements as well as for revolutionary developments. Such tasks may be for instance, the continuous sensing of environmental parameters in the transportation of fresh goods and pharmaceuticals or the integration of tags into (and not onto) goods and packages in order to create truly "smart objects". Other examples are the building of decentralised networks of smart tags in environmental monitoring systems, security and privacy enhancing technologies for health, public transportation and retail applications and security architectures for the protection of sensitive information in networks which are open to many participants.

The according technology roadmap for RFID opens up a great number of R&D areas covering a wide range of technology fields. There are topics from microelectronics, micromechanical systems, systems design, RF technology, new materials, i.e. non-silicon, cryptography and ICT architectures. Radio frequency identification cannot be reduced to a single technology which will pave the further way as a major breakthrough. On the contrary, R&D for RFID systems needs a crossover approach which puts its focus not only on long-term challenges such as polymer computing, but also on pragmatic but essential issues like antenna design.

R&D Environment

An important issue for an effective research and development policy is to create an open innovation environment. It must therefore be ensured that the close cooperation between research institutions, large companies and also SMEs is facilitated. This close cooperation is one prerequisite for a rapid flow of ideas up- and down the R&D chain. The analysis has shown that the particular focus, emphasis, or support concerning RFID technology and application differ widely across Europe. New Member States and Central and Eastern European countries seem to lag behind, especially in pre-competitive applied R&D programmes. In order to reap the benefits of this technology, a Europe-wide balance should be aimed at.

Further a balanced support of R&D themes is needed. This does not only aim at the topics that are supported themselves, but also at how support should be given to be most efficient. Here it is important to support clustering and exchange of information between different projects, to foster both regional and thematic clustering and to encourage cooperation driven by research capabilities of actors and not by institutional characteristics such as company size or country. This could be achieved by creating an open innovation environment, where research, application development, prototyping and market introduction not only co-exist, but exhibit intensive interrelations promoting quick transition from research to application, rapid and effective testing of novel ideas, and backflow of market and application experience to research.

RFID research policy should not only be directed to the middle- and long-term topics but also support the "go to market" step of novel applications. Europe should take the initiative to promote RFID applications in areas of societal importance e.g. drug authentication, efficient logistics and transport, automotive industry or transnational use of eGovernance techniques by supporting the application of RFID-based solutions. State procurement should open up for novel, potentially risky solutions to foster technology development and application areas where RFID technology can provide answers.

An increased level of public awareness and acceptance of RFID can be achieved by supporting programs which deal with the issue of privacy and data security. By acting as lead customer for applications of public interest in the fields of healthcare, security, or environmental management, Europe can integrate RFID into public services, thereby familiarising European citizens with RFID.

8 Conclusion: The Next Steps for Europe

8.2 The Stakeholder Perspective

In the following section, all recommendations developed within this book have been summarised and depicted for each stakeholder group. In this way, there is a good overview of the fields in which the different stakeholders should play a role in supporting the further development of RFID in Europe and start taking the next steps with commitment. Recommendations listed in the below tables often relate to more than one stakeholder group. The "Support creating fewer but broader accepted standards" recommendation, for example, is aimed at several stakeholder groups; obviously the "Standardisation Organisations" group is addressed, but also "Business Associations", "RFID Technology Suppliers" and "RFID End User Companies" are addressed by this recommendation. Technology suppliers and user companies for example can either indirectly propose their positions via the specific business associations to put forward their ideas on standards or they can play a direct and active role in the standardisation organisations.

Another example is the "Involve relevant stakeholders in the process of updating the RFID Implementation Checklist as guiding instrument for writing guidelines" recommendation. This applies to all stakeholder groups, as all stakeholders should be included in this process. However, only the business organisations are involved actively, because they should contact the different stakeholders making sure to integrate their informational needs. It is important to remember when looking at this overview of recommendations that each listed recommendation may not apply explicitly to a stakeholder group, but the group will certainly have an effect on the development of the technology in this respect. To make it easy for the reader, the respective chapter and page number is added so that the role of each stakeholder concerning the different recommendation can easily be found in the chapters in detail.

Research and Development

The stakeholder group "Research and Development" should be involved in the discussions about standards and guidelines, and legal aspects whenever it comes to technological topics to use their expertise. In addition, researchers should actively enhance cooperation with other institutes to improve information flow and in this way support Europe-wide RFID deployment.

Recommendation	Chapter, Page
Support basic standards like frequency allocations, air interface protocols and tag data specifications	3, 20
Eliminate the necessity for listen-before-talk	3, 26
Become actively involved in the process of updating the RFID Implementation Checklist as a guiding instrument for writing guidelines	4, 79
Privacy by design should be promoted	5, 99
Assess the environmental impact of RFID	5, 102
Conduct research aiming at minimising environmental effects of RFID	5, 103
Create and support an open innovation environment to ease the rapid flow of ideas through the RFID value chain	7, 169
Support Europe-wide RFID deployment	7, 170

8 Conclusion: The Next Steps for Europe

RFID Technology Suppliers

The stakeholder group "RFID Technology Suppliers" should be actively involved in the discussions about standards and guidelines and legal aspects especially when it comes to technological and system implementation topics. Further they should cooperate closely with researchers to improve the information flow and especially to shorten the time from "innovation" until "go to market".

Recommendation	*Chapter, Page*
Support creating fewer but broader accepted standards	3, 19
Support basic standards like frequency allocations, air interface protocols and tag data specifications	3, 20
Support the topic of providing additional UHF spectrum	3, 25
Support elimination of the necessity for listen-before-talk	3, 26
Support the standardisation process in Europe	3, 30
Identify the needs for data protection and data security of different application fields and become actively involved in developing corresponding guidelines and suitable security standards	3, 44
Support the theme that guidelines have to be application- and addressee-specific	4, 79
Support the process of updating the RFID Implementation Checklist as guiding instrument for writing guidelines	4, 79
Privacy by design should be promoted	5, 99
Self regulation and guidelines enacted by industry should be welcome	5, 99
Ensure compliance with EMF legal framework	5, 101
Assess the environmental impact of RFID	5, 102
Work towards a community patent for Europe	5, 114
Create and support an open innovation environment to ease the rapid flow of ideas through the RFID value chain	7, 169
Support Europe-wide RFID deployment	7, 170

Business Associations

The stakeholder group "Business Associations" should initiate and actively participate in the discussions about standards, guidelines, legal aspects, and research policy to put forth the topics of the companies they represent. They should cooperate closely with other representatives of their stakeholder group and also with all other stakeholder groups to ensure that urgent topics are spread and discussed quickly.

Recommendation	Chapter, Page
Support creating fewer but broader accepted standards	3, 19
Support the topic of providing additional UHF spectrum	3, 25
Ensure cooperation between different code issuing agencies	3, 27
Harmonise the standardisation for Internet of Things and RFID	3, 28
Encourage European companies to support the standardisation process	3, 30
Support the development of data exchange standards	3, 31
Harmonised international standards and application specific standards should only be established if generic standards are not sufficient	3, 40
Identify the needs for data protection and data security of different application fields and develop corresponding guidelines and suitable security standards	3, 44
When drafting guidelines they have to be application- and addressee-specific	4, 79
Involve relevant stakeholders in the process of updating the RFID Implementation Checklist as guiding instrument for writing guidelines	4, 79
Foster the use and the continuous development of the RFID Implementation Checklist	4, 80
Public information on RFID is crucial	5, 97
Self regulation and guidelines enacted by industry should be welcome	5, 99
Monitor RFID developments to ensure compliance with EMF legal framework	5, 101
Support work towards a community patent for Europe	5, 114
Encourage European participation in patent pools	5, 114
Follow an international approach with regards to Intellectual Property Rights	5, 114
Support the idea of an open innovation environment to ease the rapid flow of ideas through the RFID value chain	7, 169
Support Europe-wide RFID deployment	7, 170

8 Conclusion: The Next Steps for Europe

Government and Governmental Institutions

The stakeholder group "Government and Governmental Institutions" should support and follow the discussions about standards, guidelines, legal aspects, and research policy and listen to the recommendations of all stakeholder groups in order to establish a policy framework that supports the needs of the different stakeholder groups in the best way possible.

Recommendation	Chapter, Page
Support basic standards like frequency allocations, air interface protocols and tag data specifications	3, 20
Support the development of royalty-free standards	3, 20
Provide additional UHF spectrum	3, 25
Support the elimination of the necessity for listen-before-talk	3, 26
Harmonised international standards and application specific standards should only be established if generic standards are not sufficient	3, 40
Identify the needs for data protection and data security of different application fields and develop corresponding guidelines and suitable security standards	3, 44
Support the approach of guidelines have to be application- and addressee-specific	4, 79
Become actively involved in the process of updating the RFID Implementation Checklist as guiding instrument for writing guidelines	4, 79
Foster the use and the continuous development of the RFID Implementation Checklist	4, 80
Ensure an appropriate radio spectrum framework	5, 108
The existing legal framework regarding data protection is sufficient for RFID technology applications, the focus should be put on the enforcement	5, 97
Public information on RFID is crucial	5, 97
The protection of personal data should be ensured without imposing unreasonable burdens upon the RFID users	5, 98
Self regulation and guidelines enacted by industry should be welcome	5, 99
Monitor RFID developments to ensure compliance with EMF legal framework	5, 101
Other means aside from CE marking to inform and protect citizens should be enacted	5, 101
Assess the environmental impact of RFID	5, 102
Encourage European participation in patent pools	5, 114
Follow an international approach with regards to IPR	5, 114
Create and support an open innovation environment to ease the rapid flow of ideas through the RFID value chain	7, 169
Support Europe-wide RFID deployment	7, 170
Provide balanced support for R&D themes and topics	7, 170
Support "go to market" of novel applications	7, 173

Standardisation Organisations

The stakeholder group "Standardisation Organisations" should initiate and actively raise their voice in the discussions about all topics concerning RFID standards directly and indirectly. Further, they should cooperate closely with other Standardisation Organisations and allow all interested parties to participate in the standardisation process by offering easy and affordable access, e.g. also for SMEs.

Recommendation	Chapter, Page
Support creating fewer but broadly accepted standards	3, 19
Support basic standards like frequency allocations, air interface protocols and tag data specifications	3, 20
Support the development of royalty-free standards	3, 20
Support the topic of providing additional UHF spectrum	3, 25
Eliminate the necessity for listen-before-talk	3, 26
Ensure cooperation between different code issuing agencies	3, 27
Harmonise the standardisation for Internet of Things and RFID	3, 28
Encourage European companies to support the standardisation process	3, 30
Support the development of proper data exchange standards	3, 31
Harmonised international standards and application specific standards should only be established if generic standards are not sufficient	3, 40
Work towards international cooperation of RFID standardisation organisations	3, 40
Identify the needs for data protection and data security of different application fields and develop corresponding guidelines and suitable security standards	3, 44
Support the approach that guidelines have to be application- and addressee-specific	4, 79
Become actively involved in the process of updating the RFID Implementation Checklist as guiding instrument for writing guidelines	4, 79
Foster the use and the continuous development of the RFID Implementation Checklist	4, 80
Support the discussion on an appropriate radio spectrum framework	5, 108
Privacy by design should be promoted	5, 99
Work towards a community patent for Europe	5, 114
Encourage European participation in patent pools	5, 114
Follow an international approach with regards to Intellectual Property Rights	5, 114
Support an open innovation environment to ease the rapid flow of ideas through the RFID value chain	7, 169

Quasi-Autonomous and Non-Governmental Organisations

The stakeholder group "Quasi-Autonomous and Non-Governmental Organisations" should actively raise their voice in the discussions about RFID when it comes to topics interesting for the groups they represent. They should cooperate closely with other representatives of their stakeholder group and also with all other stakeholder groups to ensure that urgent topics are spread and discussed quickly.

Recommendation	*Chapter, Page*
Support the development of proper data exchange standards	3, 19
Harmonised international standards and application-specific standards should only be established if generic standards are not sufficient	3, 40
Identify the needs for data protection and data security of different application fields and support the development of corresponding guidelines and suitable security standards	3, 44
Be actively involved in the process of updating the RFID Implementation Checklist as a guiding instrument for writing guidelines	4, 79
The protection of personal data should be ensured without imposing unreasonable burdens upon the RFID users	5, 98
Other means aside from CE marking to inform and protect citizens should be enacted	5, 101
Support the assessment of environmental impact of RFID	5, 102
Encourage European participation in patent pools	5, 114
Support an international approach with regards to Intellectual Property Rights	5, 114
Support an open innovation environment to ease the rapid flow of ideas through the RFID value chain	7, 169

RFID End User Companies

The stakeholder group "RFID End User Companies" should actively put forth their opinion in the discussions about standards, guidelines, legal aspects, and research policy especially whenever it comes to application topics. It is important that they raise their voice individually or via their business associations in order to actively shape the RFID future.

Recommendation	*Chapter, Page*
Support creating fewer but broadly accepted standards	3, 19
Support the topic of better cooperation between different code issuing agencies	3, 27
Support harmonising the standardisation for Internet of Things and RFID	3, 28
Support the development of proper data exchange standards	3, 31
Harmonised international standards and application-specific standards should only be established if generic standards are not sufficient	3, 40
Identify the needs for data protection and data security of different application fields and support the development corresponding guidelines and suitable security standards	3, 44
Support the approach that guidelines have to be application- and addressee-specific	4, 79
Become actively involved in the process of updating the RFID Implementation Checklist as guiding instrument for writing guidelines	4, 79
Public information on RFID is crucial	5, 97
Self regulation and guidelines enacted by industry should be welcome and used	5, 99
Ensure compliance with EMF legal framework	5, 101
Support work towards a community patent for Europe	5, 114
Support an open innovation environment to ease the rapid flow of ideas through the RFID value chain	7, 169
Support Europe-wide RFID deployment	7, 170

References

Adcock C (2007) A Global, User Driven Perspective on Standardisation, Presentation held at the conference "On RFID: Next steps towards the Internet of Things". Lisbon, November 16, 2007. http://www.rfid-outlook.pt/pdf/ChrisAdcock.pdf Accessed 23rd April 2008

Aho E (2008) Creating an Innovative Europe. Report of the Independent Expert Group on R&D and Innovation appointed and following the Hampton Court Summit and chaired by Mr. Esko Aho. EUR 22005, Luxembourg, January 2006: European Communities

Alien Technology Corp (2008) FSA Manufacturing. http://www.alientechnology.com/fsa_manufacturing.php Accessed 18th May 2008

Article 29 Data Protection Working Party (2005) Working document on data protection issues related to RFID technology, 10107/05/EN WP 105 January 19, 2005

Article 29 Data Protection Working Party (2007) Opinion 4/2007 on the concept of personal data, 01248/07/EN WP 136, adopted on 20th June

Barua A, Mani D, Whinston A (2006) Assessing the Financial Impacts of RFID Technologies on the Retail and Healthcare Sectors, Austin: University of Texas

Bizer J, Spiekermann S, Günther O (2006) TAUCIS Technikfolgenabschätzung. Ubiquitäres Computing und Informationelle Selbstbestimmung – On behalf of the German Federal Ministry of Education and Research, Independent Center for Privacy Protection Schleswig-Holstein and Institute of Information Systems at Humboldt University: Berlin

BMBF (2004) Framework Programm 2004–2009 Mikrosysteme, BMBF (German Federal Ministry of Education and Research)
http://www.bmbf.de/pub/mikrosysteme.pdf
Accessed 23rd April 2008

BMBF (2007a) Technologieintegrierte Datensicherheit bei RFID Systemen. BMBF (German Federal Ministry of Education and Research).
http://www.sit.fraunhofer.de/fhg/Images/RFID-Studie2007_tcm105-97982.pdf
Accessed 23rd April 2008

BMBF (2007b) IKT 2020, Forschung für Innovationen. BMBF (German Federal Ministry of Education and Research). http://www.bmbf.de/pub/ikt2020.pdf
Accessed 23rd April 2008

BMWi (2007) European Policy Outlook RFID, Berlin.
http://www.nextgenerationmedia.de/Nextgenerationmedia/Navigation/en/rfid-conference,did=169936.html
Accessed 16th April 2008

Bovenschulte M, Gabriel P, Gassner K, Seidel U (2007) RFID: Prospectives for Germany. The state of radio frequency identification-based applications and their outlook in national and international markets. Berlin: Federal Ministry of Economics and Technology

BRIDGE (2007) High level design for Discovery Services.
http://www.bridge-project.eu/data/File/BRIDGEWP02HighleveldesignDiscovery Services.pdf
Accessed 6th May 2008

BSI (2005) Security Aspects and Prospective Applications of RFID Systems. BSI(Federal Office for Information Security)
http://www.bsi.de/fachthem/rfid/RIKCHA_englisch.pdf
Accessed 19th May 2008

BSI (2008) Technischen Richtlinien für den sicheren RFID-Einsatz. Draft version.
http://www.bsi.bund.de/presse/pressinf/221107_rfidworkshop.htm
Accessed 9th January 2008

Buckley J (ed.) (2006) Conference Report: From RFID to the Internet of Things, Pervasive Networked Systems. Brussels: CCAB

Bullinger, H-J, ten Hompel, M (2008) (eds.) Internet der Dinge. Berlin/Heidelberg/New York: Springer

CDTI (2008) CDTI.es. www.cdti.es.
Accessed 28th April 2008

References

CERP (2007) Cluster of European RFID Projects [CERP]: Working Paper on Future RFID Research Needs.
http://www.rfid-in-action.eu/cerp/cerp-working-paper/working-paper-on-future-rfid-research-needs
Accessed 16th April 2008

CIP (2007) Competitiveness and Innovation Framework Programme: ICT Policy Support Programme, work programme 2007.
http://ec.europa.eu/information_society/activities/ict_psp/library/ref_docs/index_en.htm
Accessed 6th May 2008

COM(2000)199 final: Report from the Commission to the Council, the European Parliament, and the Economic and Social Committee on the implementation and effects of Directive 91/250/EEC on the legal protection of computer programs

COM(2006)181 final: Towards a Global Partnership in the Information Society: Follow-up to the Tunis Phase of the World Summit on the Information Society (WSIS [World Summit on the Information Society])

COM(2007)165 final: Enhancing the patent system in Europe

COM(2007)96 final: Radio Frequency Identification (RFID) in Europe: steps towards a policy framework

COM(2007)700 final: Reaping the full benefits of the digital dividend in Europe: A common approach to the use of the spectrum released by the digital switchover, p5

COM(2007) 799 final: Pre-commercial Procurement: Driving innovation to ensure sustainable high quality public services in Europe. Example of a possible approach for procuring R&D services applying risk-benefit sharing at market conditions, i.e. pre-commercial procurement.
Commission staff working document

D'Aveni R A, MacMillan I C (1990) Crisis and content of managerial communications: A study of the focus of attention of top managers in surviving and failing firms.
Administrative Science Quarterly, 35: p634–657

Daymon C, Holloway I (2002) Qualitative research methods. Malta: Gutenberg Press

Deffner, G (1986) Microcomputers as aids in Gottschalk-Gleser rating.
Psychiatry Research, 18: p151–159

DG INFSO (2006) RFID portfolio of European Research,
Factsheet 54, October 2006: DG INFSO

Dienel, P (2002) Die Planungszelle. Wiesbaden: Westdeutscher Verlag

Environment DG (2007) Waste electrical and electronic equipment.
http://ec.europa.eu/environment/waste/weee/index_en.htm
Accessed 6th May 2008

EPCglobal (2003) EPCglobal Intellectual Property Policy Working Group Declaration.
http://www.epcglobalinc.org/what/ip_policy/031223EPCgloballPPolicy12152003A.pdf
Accessed 6th May 2008

EPCglobal (2007) EPCglobal Intellectual Property Policy FAQ's.
http://www.epcglobalinc.org/what/ip_policy/EPCglobal_IP_FAQ_FINAL_Feb_2007_V13.pdf
Accessed 30th April 2008

EPoSS (2007) European Technology Platform on Smart Systems Integration [EPoSS]: Implementing the European Research Area for Smart Systems Technologies. Strategic Research Agenda v1.3.
http://www.smart-systems-integration.org/public/documents/EPoSS_publications/EPoSS_SRA_1_3
Accessed 16th April 2008

Eschinger C (2008a) Market Trends: Radio Frequency Identification, Worldwide, 2007–2012. Gartner, 11th February 2008, ID Number: G00154771 in: Sullivan, L (2008) Gartner Research: RFID 2008 Market Forecast To Hit $ 1.2 Billion.
RFID World, http://www.rfid-world.com/news/206900115
Accessed 26th May 2008

Eschinger C (2008b) Market Trends: Radio Frequency Identification, Worldwide, 2007–2012. Gartner, 11th February 2008, ID Number: G00154771 in: Computing SA (2008) Global RFID revenue to surpass $ 1,2bn in 2008 – Gartner.
Computing SA. http://www.computingsa.co.za/article.aspx?id=713786
Accessed 26th May 2008

ETAG (2007) Policy Options for the Improvement of the European Patent System.
http://www.europarl.europa.eu/stoa/publications/studies/stoa16_en.pdf
Accessed 30th April 2008

ETSI (2008) European Telecommunications Standards Institute (ETSI) Technical Report Draft ETSI TR 102 649-2 v1.1.1_0.15(2008-02).
France: Sophia Antipolis, Cedex

EU (2008) EU Tube – RFID: The future begins now.
http://www.youtube.com/watch?v=Zg1VKJUrxi4
Accessed 30th April 2008

European Commission (2000a) The Economic Impact of a Patentability of Computer Programs.
http://ec.europa.eu/internal_market/indprop/docs/comp/study_en.pdf
Accessed 30th April 2008

European Commission (2000b) Commission proposes the creation of a Community Patent, IP/00/714.
http://europa.eu/rapid/pressReleasesAction.do?reference=IP/00/714&format=HTML&aged=1&language=EN&guiLanguage=fr
Accessed 30th April 2008

European Commission (2005) DG Internal Market and Services Working Paper, First evaluation of Directive 96/9/EC on the legal protection of databases, December 12, 2005.
http://ec.europa.eu/internal_market/copyright/docs/databases/evaluation_report_en.pdf
Accessed April 30th 2008

European Commission (2006a) Future Patent Policy in Europe, Preliminary Results of the Public Consultation

European Commission (2006b) Future Patent Policy in Europe, Report on the Public Hearing, Brussels, July 12, 2006.
http://ec.europa.eu/internal_market/indprop/docs/patent/hearing/report_en.pdf
Accessed 30th April 2008

European Commission (2007) Communication from the Commission to the European Parliament, the Council, the European Economic and Social Committee and the Committee of the Regions: Radio Frequency Identification (RFID) *in*
European Commission (2008) Directorate General Environment website, "Waste Electrical and Electronic Equipment".
http://ec.europa.eu/environment/waste/weee/index_en.htm
Accessed 30th April 2008

European Communities (ed.) (2008) On RFID: The next step to the Internet of Things, p269, Brussels 2008

European Parliament STOA (2006) RFID and Identity Management in Everyday Life. IPOL /A/STOA/2006-22

European Parliament (2008) Resolution on the second Internet Governance Forum, held in Rio de Janeiro from 12 to 15 November 2007, B6-0041/2008.
http://www.europarl.europa.eu/sides/getDoc.do?pubRef=-//EP//TEXT+MOTION+B6-2008-0041+0+DOC+XML+V0//EN
Accessed 6th May 2008

Fabian B, Spiekermann S (no date) Security Analysis of the Object Name Service (ONS) for RFID (Paper written within the TAUCIS Study)
http://www.taucis.hu-berlin.de/_download/security_analysis.pdf
Accessed 30th April 2008

Finkenzeller K (2003) RFID-Handbook, Fundamentals and Applications in Contactless Smart Cards and Identification 2nd edition, New York: Wiley

Fleisch E, Mattern F (2005) Das Internet der Dinge: Ubiquitous Computing und RFID in der Praxis. Berlin, Heidelberg: Springer

Fraunhofer IPT (2008) German companies do not tap full potential of RFID. Fraunhofer Institut of Production Technology (IPT).
http://www.ipt.fraunhofer.de/EN/press/StudieRFID.jsp
Accessed 15th April 2008

Gampl B, Lange S, Holweger M, Melià J, Uestuen A (2008a) Analysis of Guidelines for RFID Implementation: Status and Recommendations.
CE RFID, http://www.rfid-in-action.eu
Accessed 14th April 2008

Gampl B, Robeck M, Clasen M (2008b) The RFID Reference Model. In: Unternehmens-IT: Führungsinstrument oder Verwaltungsbürde? (Conference Proceedings). Bonn: Gesellschaft für Informatik

GCI (2003) Global Commerce Initiative: EPC Roadmap.
http://www.gs1-germany.de/content/e39/e466/e468/datei/ccg/gci_epc_roadmap.pdf
Accessed 16th April 2008

Gliesche M, Helmigh, M (2008) Impact of RFID Mass Deployment on Waste and Recycling Systems, Dortmund University, Logistics Dept., supported by the German Federal Ministry for Education and Research.
http://www.flog.mb.uni-dortmund.de/forschung/download/StudieAuswirkungeneinesRFIDMasseneinsatzes.pdf (German)
Accessed 16th May 2008

Goel R (2007) Managing RFID Consumer Privacy and Implementation Barriers. Information Systems Security, 16:p217–223

References

GS1 France (2007): Des codes & des news, Newsletter, December 2007.
http://www.gs1news.fr/NL/NL046_uk.cfm
Accessed 30th April 2008

GS1 France (2008): Des codes & des news, Newsletter, January 2008.
http://www.gs1news.fr/NL/NL047_UK.cfm
Accessed 30th April 2008

Harrop P, Das R (2006) RFID Forecasts, Players and Opportunities 2006 to 2016. Cambridge: IdTechEx, p5–9

Holznagel B, Bonnekoh M (2006) RFID: Rechtliche Dimensionen der Radiofrequenz-Identifikation, Berlin: Informationsforum RFID

ICANN (2008) Response to the Midterm Review of the Joint Project Agreement, 9 January, 2008,
http://www.icann.org/correspondence/dengate-thrush-to-sene-09jan08.pdf
Accessed 30th April 2008

International Labour Organization Geneva (1997) Protection of Worker's Personal Data – An ILO Code of Practice, 4 February 2002,
http://www.ilo.org/public/english/protection/safework/cops/english/download/e000011.pdf
Accessed 30th April 2008

Internet Governance Forum (2008) Multistakeholder Advisory Group (MAG), Meeting Summary, February 27–28, 2008, Geneva.
http://www.intgovforum.org/Feb_igf_meeting/MAG.Summary.28.02.2008.v2.pdf
Accessed 30th April 2008

IPTS (2007) RFID technologies: Emerging Issues, Challenges and Policy Options. European Commission.
http://ftp.jrc.es/eur22770en.pdf
Accessed 23rd April 2008

Kennisportal.com (2008) RFID. TNO Kennisportal-RFID.
http://www.kennisportal.com/main.asp?ChapterID=3558
Accessed 23rd April 2008

Kruse A, Mortera-Martinez C, Corduant V, Lange S (2008) Recommendations for further European Legislative Activities.
CE RFID, http://www.rfid-in-action.eu
Accessed 14th April 2008

Maghiros I, Rotter P, van Lieshout, M (ed.) (2007) RFID Technologies: Emerging Issues, Challenges and Policy Options. Institute for Prospective Technological Studies,Luxembourg: Office for Official Publications of the European Communities 2007

Ministère de l'Economie (2008) Pôles de Compétitivité.
http://www.competitivite.gouv.fr/
Accessed 23rd April 2008

Ministerie van Economische Zaken (2006) RFID in Nederland.
http://www.minez.nl/dsresource?objectid=145154&type=PDF
Accessed 23rd April 2008

Morris R (1994) Computerized Content Analysis in Management Research: A Demonstration of Advantages & Limitations.
Journal of Management, 20(903), p903–931

Netzwerk elektronischer Geschäftsverkehr (2008) Interaktive Checkliste zu RFID. Initiative "Netzwerk Elektronischer Geschäftsverkehr"
http://www.ec-net.de/EC-Net/Navigation/Service/impressum.html
Accessed 16th April 2008

OECD (2006) Directorate for Science, Technology and Industry-Committee for Information, Computer and Communications Policy: Radio Frequency Identification (RFID): Drivers, Challenges and Public Policy Considerations.
DSTI/ICCP (2005)19/
FINAL, Paris: OECD, 27 February 2006

OECD (2008) OECD Working Party on the Information Economy: RFID Applications, Impacts and Country Initiatives.
DSTI/ICCP/IE(2007)13/
REV1, Paris: OECD, 5–7 March 2008

OECD (2008a) OECD Working Party on Information Security and Privacy: Radio Frequency Identification (RFID): A Focus on Information Security and Privacy.
DSTI/ICCP/REG(2007)9/
FINAL, Paris: OECD, 14 January 2008

Office of the Privacy Commissioner of Canada (2008) Radio Frequency Identification in the Workplace: Recommendations for Good Practices. A Consultation Paper.
http://www.privcom.gc.ca/information/pub/rfid_e.pdf
Accessed 19th May 2008.

References

Pauvre N (2008) Supporting Standardisation of an Open Governance for the EPC global Network – The French Initiative to put a Theory into Practice *in* Michahelles F (ed.) (2008) Internet of Things 08, Adjunct Proceedings, St. Gallen 2008, p22–23
http://www.iot2008.org/adjunctproceedings.pdf
Accessed 30th April 2008

RFID Platform Nederland (2008) RFID Platform Nederland.
http://www.rfidnederland.nl.
Accessed 23rd April 2008

Rogers E M (1962) Diffusion of Innovations. New York: The Free Press

Rogers E M (1983) Diffusion of Innovations. USA: Macmillan *in* Inagaki N, (2007) Communicating the Impact of Communication for Development: Recent Trends in Empirical Research. World Bank Publications.
http://publications.worldbank.org
Accessed 18th March 2008

Roure F, Gorichon J, Sartorius E (2005) CGTI – Les Technologies de Radio- identification (RFID): enjeux industriels et questions socientals,
rapport II-B.9, January 2005

Schumpeter J A (1939) Business Cycles: A Theoretical, Historical, and Statistical Analysis of the Capitalist Process. 1st ed.
New York and London: McGraw-Hill

SEC(2007)312: Radio Frequency Identification (RFID) in Europe: steps towards a policy framework

Shapiro G, Markoff G (1997) In C. W. Roberts (Ed.), Text analysis for the social sciences: Methods for drawing statistical inferences from text and transcripts. Mahwah, NJ
Lawrence Erlbaum Associates, p9–31

Sicher im Netz (2008) Deutschland sicher im Netz. Deutschland sicher im Netz e.V.
http://www.sicher-im-netz.de
Accessed 16th April 2008

Sieker B, Ladkin P, Hennig J (2005) Privacy Checklist for Privacy Enhancing Technology Concepts for RFID Technology Revisited
Bielefeld: University of Bielefeld, RVS (Research Group on Networks and Distributed Systems)

Strüker J, Gille D, Faupel T (2008) RFID Report 2008 – Optimierung von Geschäftsprozessen in Deutschland. Albert-Ludwigs-Universität Freiburg: IIG-Telematik, Düsseldorf: VDI Nachrichten

TEKES (2008) Homepage TEKES Agency.
http://www.tekes.fi/eng/
Accessed 28th April 2008

Van de Voort M, Ligtvoet A (2006) Towards an RFID policy for Europe, Workshop Report, European Commission.
http://www.rfidconsultation.eu/docs/ficheiros/RFID_Workshop_Reports_Final.pdf
Accessed 23rd April 2008

Van Pottelsberghe de la Potterie B, François D (2006) The Cost Factor in Patent Systems, Working Paper WP-CEB 06-002.
Brussels: Université Libre de Bruxelles.

Verheugen G (2007) Speech at the European Patent Forum in Munich, April 20, 2007.
http://www.epo.org/topics/news/2007/070420.html
Accessed 30th April 2008

Walk E, Büth D, Desch M, Rödig M, Neubauer F, Gauby A, Hoisl A (2008) RFID Standardisation.
CE RFID, http://www.rfid-in-action.eu
Accessed 14th April 2008

Weber R (1990) Basis content analysis (2nd Ed.).
Sage Publications, CA: Thousand Oaks

Weiser M (1991) The Computer for the 21st Century.
Scientific American, 265 (3), p94–104, 1001.
http://www.ubiq.com/hypertext/weiser/SciAmDraft 3.html
Accessed 29th March 2008

WGIG (2005) Report from the Working Group of Internet Governance, Document WSIS-II/PC-3/DOC/5-E,
http://www.itu.int/wsis/docs2/pc3/off5.pdf
Accessed 30th April 2008

Wiebking L, Metz G, Korpela M, Nikkanen M, Penttilä K (2008) RFID Roadmap.
CE RFID, http://www.rfid-in-action.eu
Accessed 14th April 2008

Wilkinson R (2005) Public Procurement for Research and Innovation. Developing procurement practices favourable to R&D and innovation.
Expert Group Report EUR 21793 EN.
European Commission: Luxembourg, January 2006

Index

A

Access control system 31, 38
    Personal tracking 31, 38
    Rental systems 31, 38
ADS 118
AFNOR 17
Ahold 121
AIAG 35
AII 149
Air interface 7, 18, 20, 21, 22, 23, 24, 27, 34, 43, 74, 126, 127, 132–134, 175, 180, 181, 183, 184
    Protocol 18, 20, 22, 27, 74, 180, 181, 183, 184
Alien Technology 112, 132
Ambient intelligence 138
ANR 149
Antenna 22, 27, 123, 124, 126, 129, 132, 134–136, 171, 178
    Design 22, 132, 171, 178
    Diversity 126
Anti-collision 123, 135
ARTEMIS 172, 173
Article 29 Working Party 82, 86
Auto-ID 30, 121
AWID 112

B

Barcode 18, 27, 81, 117
Benhamou, Bernard 118
BMBF 57, 128, 146, 148
BMWi 2, 3, 54, 108, 146–148
BSI 17, 42, 79, 94, 128, 148
BSIK 152
Business association 12, 51, 64, 179, 182, 186

C

Carrefour 121
CCTV 94, 95
CDTI 153
CE 3, 5, 16, 31, 51, 53, 78, 100, 101, 175, 183, 185
CEE 170
CEN 18
CENELEC 19
Centres of Excellence 171, 173
CERP 128, 164, 167, 172
Checklist See RFID Implementation Checklist
Chip 22, 43, 91, 124, 129, 130, 132, 135, 136, 156, 171
    Coil on 132
    Design 129
CII 109
CIP 170
Code issuing agencies 27, 182, 184, 186
Commission Decision 2006/804/EC 23, 107
Contract Research Organisations 158
Council Directive
    87/372/EEC (GSM Directive) 25, 104, 106
    91/250/EEC (Legal Protection of Computer Programs) 110
Cryptography 44, 127, 137, 178

D

Data processing 27, 41, 131
Data Protection Directive 82, 83, 84, 85, 86, 87, 88, 89, 91, 92–94, 97, 177
Data quality 85

197

Data subject 81, 85, 86, 88, 89, 92, 93, 94, 97–99
Delhaize 121
Department of Trade and Industry, UK 151
DFG 146, 147
DGE 149
DIN 17, 35
Directives of the European Parliament
    2001/29/EC (EU Copyright Directive) 111
Directives of the European Parliament and of the Council
    2002/20/EC (Authorisation Directive) 104
    2002/21/EC (Framework Directive) 83
    2002/58/EC (ePrivacy Directive) 42, 83
    2002/95/EC (RoHS Directive) 102
    2002/96/EC (WEEE Directive) 102
    2004/40/EC (Electromagnetic Fields Directive) 100
    95/46/EC (Data Protection) 42, 83
    96/9/EC (Legal Protection of Databases) 110
Display
    Bi-stable 136
DNS 116, 117
Dutch Ministry of Finance 152
DWD 21

E

EAN 18, 19
ECJ 109
EFTA 17, 19
eGovernance 119, 178
eGovernment 29, 127, 174
Electronic article surveillance 36
EMC 24, 25
EMF 101, 181–183, 186
Energy See Power
Environment 3, 4, 7, 17, 23, 41, 63, 81, 83, 93, 98, 101, 102, 103, 111, 116, 117, 123, 127, 133, 153, 168, 169, 171, 174, 178, 180, 181, 182, 183, 184–186
Environmental protection 6, 176
EPC 18, 19, 23, 27, 29, 30, 41, 44, 61, 109, 117, 118, 128, 134
EPCglobal 12, 18, 19, 22, 27, 29, 30–32, 42, 61, 80, 112, 114, 117, 118, 120, 128
    Generation 2 See Gen2E–9, 117
EPO 108, 109
EPoSS 128, 164, 172

ERM 22–24
ERP 23, 24
ESDS 118
ETSI 17, 19, 21, 22, 23, 24–26
EUREKA 141, 144, 148, 149, 153, 157, 158, 159–161, 165, 166
European Biometric Passport 91
European Patent Convention 108, 114
European Policy Outlook 54, 108, 120

F

FFG 154
FP5 144, 163
FP6 158, 162, 163
FP7 163, 164, 170
Fraunhofer Gesellschaft 146, 147
FUB 153
FWF 154

G

GCI 128
Gen2 1, 18, 22, 23, 44, 112, 128
GHz 24, 25, 100, 106, 126, 133
GLN 27
Governance 116
    Internet 115, 116, 119, 120, 177
GRAI 27
GSM 104
GTIN 27
Guideline 37, 42, 48, 49, 50, 51, 52, 53, 54, 55, 56, 57, 58, 59, 60, 61, 62, 63, 65, 66, 67, 68, 69, 71, 72, 77–79

H

Harmonisation
    Patent law 109
    Radio regulation 20, 23, 104, 106
Health 2, 3, 6, 7, 11, 17, 41, 49, 50, 57, 76, 81, 85, 100, 101, 122, 175, 177
Healthcare 1, 11, 57, 58, 96, 152, 168, 170, 175, 178
HF 18, 22, 36, 44, 123, 125, 126, 129, 132, 134, 136

I

IATA 33, 37
IC See Chip
ICANN 115–117
ICAO 42, 113

Index

ICT 1, 3, 17, 82, 94, 116, 119, 138, 141, 147, 156, 170, 175, 178
ICT architecture 138, 178
Identification 1, 27, 57, 62, 63, 82, 87, 95, 104, 112, 126
  Indirect 87, 89
Identifier 27, 43, 44, 117
IEEE 35, 41
Innovator 111
Intellectual Property Rights 7, 19, 20, 81, 108, 109, 110, 112, 113, 114, 115, 169, 177, 182, 183–185
  Policy 108
Internet 3, 28, 29, 30, 31, 43, 47, 53, 114, 115, 116, 117, 118–120, 122, 127, 177, 182, 184, 186
Internet of Things 3, 28, 29, 30, 31, 114, 115, 116, 117, 118–120, 122, 127, 182, 184, 186
Interoperability 26, 74, 115, 133, 176
Interreg 144, 161
IP 112
IPR See Intellectual Property Rights
ISM 21, 24, 25
ISO 12, 17, 18, 20, 21, 22, 23, 24, 27, 35, 38, 41, 42–44, 112, 113, 134
Item tagging 123, 124, 138
ITF 22, 23
ITU 16
ITU-R 16

K

Kill command 94
KIROLTEK 154

L

LAN 24
LF 21, 32, 123, 125, 129, 132, 134
Librfid 113
Listen-before-talk 23, 26, 180, 181, 183, 184
Logistical tracking and tracing 12, 33, 121
  Closed loop logistics 32, 33, 89
  Dangerous goods logistics 32, 33
  In-house logistics 32, 33
  Manufacturing logistics 32, 33
  Open logistics 32, 33
  Postal applications 32, 33

Logistics 1, 2, 32, 33, 36, 37, 41, 54, 62, 63, 68, 72, 89, 90, 96, 115, 121, 127, 146, 147, 152, 153, 159, 161, 162, 165–167, 170, 174, 178
London Agreement 108

M

Mass application 3, 124, 125, 138, 148
MEMS 146
METRO Group 60, 61, 121
MHz 18, 21, 22–24, 26, 107, 128, 136
MINAMI 165
Ministerie van Economische Zaken 151
Multi-hop reads 138

N

NEPTUNO 153, 154
Next Generation Media 147
NGO 12, 64, 71, 185
Non-silicon technologies 135
NORDITE 159, 167
NTIA 116, 117
NXP 136, 156

O

OECD 2, 93, 94, 96, 97
OMEDIS 153, 154
ONS 29–31, 117, 118
Open source 109, 113, 114, 172, 177
openMRTD 113
OpenPCD 113
Opt-in 92
Opt-out 92
ORFID 136
OrganicID 136
OSEO Anvar 149

P

Packaging 129
Patent
  Community Patent 109, 110, 114, 177, 181, 182, 184, 186
  European Patent Convention 108, 114
  Pool See RFID Patent Pool
PCD 113
PdC 149, 150, 165
  Logistique Seine-Normandie 150
  MINALOGIC 150
  SCS 150
  TES 150

Personal data 3, 7, 42, 43, 45, 57, 82, 83, 84, 85, 86, 87, 88, 89, 90, 91, 92, 93, 94, 95, 96, 97–99
PET 122
PISA 94
PolyIC 136
Power
    Consumption 122, 124, 125, 130, 136, 171
PRIME 165
Privacy 41, 44, 53, 57, 60, 63, 81, 82, 94–96, 99, 122, 165, 180, 181, 184
    by Design 94, 99, 143, 165, 172, 173
Product codes 27
Product safety, quality and information 36
    Customer information systems 36, 37
    Electronic goods 36, 37
    Fast moving consumer goods 36, 37
    Fresh/perishable foods 36, 37
    Pharmaceutical 37
    Textile goods 36, 37
Production, Monitoring and Maintenance 11, 12, 115
    Aeroplanes 34, 35
    Archive systems 34
    Asset management 34
    Automation/process control 34, 35
    Facility management 34
    Food and consumer goods 34, 35
    Vehicles 34, 35

Q

QAM 133
Quasi-autonomous organisation 185

R

R&D
    Environment 141, 175, 178
    Policy 8, 141, 144, 148, 159, 166–168, 173
    Programmes 141, 142, 145, 149, 158, 164, 166, 168, 170, 173, 178
Radio spectrum 22, 23, 24, 25, 103, 104, 105, 106–108, 176, 183, 184
    Framework 20, 25, 108, 176, 183, 184
RAND 113
Reader 17, 23, 27, 29, 89, 92, 123, 124, 125, 126, 127, 129, 130, 133, 134–136, 138, 143, 164, 165, 171, 179
    Multi-air-interface 126
Reader-to-tag 23
Reding, Vivian 3

REGINS-RFID 61, 153, 161
Regulative framework 175, 176
Retail 57, 60, 62
Rewe 121
RF 22, 125, 127, 129, 131, 134, 168, 171, 178
RFID
    Application fields 7, 9, 48, 55
    Consortium 112, 114
    Economic effects 2, 3
    Emergence 29, 31, 41
    End user company 12, 179, 186
    Governance 115
    Implementation checklist 7, 48, 49, 50, 51, 66, 70, 72, 78, 79, 80, 176, 179, 180, 181, 182, 183, 184–186
    Implementation guideline 54, 78
    Lab 63, 158
    Legislative framework 81
    Patent pool 111, 113, 114, 177, 182, 183–185
    Tag See Tag
    Technology roadmap 127
    Transponder See Tag
    User 2, 6, 7, 12, 28, 49, 54, 70, 77, 79, 99, 123, 124, 132, 175, 185
RFID Platform Nederland 151
RFID Reference Model 9, 11, 26, 31, 48, 54, 55, 72, 115, 143, 164
RFID Stakeholder Model 9, 12, 48
Rogers, Everett M. 4, 5, 168
ROI 5, 49, 60, 78, 124, 132, 143, 147, 148, 171

S

SAE 35
Sapienza University of Rome 153
SAW 135
School of Management of the Politecnico di Milano 153
Schumpeter, Josef 168
SD-Card 126, 134
SDO 17
Security 41, 42, 44, 73, 74, 76, 79, 112, 118, 127, 147
Self regulation 50, 75, 181–183, 186
Sensor 1, 40, 41, 121, 122, 124, 128, 131, 135, 136–138, 148, 159, 165
    Functionality 40, 41, 148
    Nano power 124, 136
    Network 122, 128, 138
    No-power 137
SenterNovem Agency 152

Index

Sleep command 94
Smart label 146
Smart object 122, 177
Smart system 11, 143, 160, 172
SME 109, 152, 171
SNR 125
SRA 163, 173
SRD 22, 24, 26
SSCC 27
Stakeholder 48, 179
Standard 16, 17, 18, 19, 20, 21, 22, 23, 24, 25, 26, 29, 32, 33, 35, 37, 38, 41, 42–44, 52, 57, 92, 105, 106, 111, 113, 117, 133, 134, 148
    Application specific 7, 31, 40, 182–184
    Data exchange standards 31, 59, 182, 184–186
    Generic 36, 38, 39, 40, 176, 182, 183, 184–186
    Packaging 40
Standardisation
    Organisation 12, 16, 17–19, 26, 40, 51, 179, 184
    Process 15, 30, 35, 39, 181, 182, 184
Standards
    Royalty-free 19, 20, 117, 183, 184
Supply chain 2, 11, 27, 29, 30, 32, 39, 57, 61, 62, 88, 118, 121, 127, 132, 153, 167, 170

T

Tag 1, 2, 3, 11, 17, 20, 21, 22, 23, 27, 29, 32, 41, 43, 44, 50, 58, 63, 74, 75, 81, 89, 91, 92, 93, 94, 97, 122, 123, 124, 125, 126, 127, 128, 129, 130, 132, 133, 135, 136–138, 143, 164, 165, 171, 176, 180, 181, 183, 184
    Active 123, 125, 128, 129, 131, 138
    Crypto 125, 137
    Inlay-based 41
    Passive 123–125, 128, 130, 135, 136, 171
    Semi-passive 123, 125
Sensor 1, 40, 121, 137
Technology Diffusion Model 4
Technology research need 175, 177
Technology supplier 12, 51, 175, 179
TEKES 157–159
Tesco 121
Traceability 34, 36, 153
Trans-Atlantic Economic Council 27
Treaty on European Union 83
TRIPS 109, 111

U

UHF 22, 23, 24, 25, 26, 34, 36, 44, 104, 107, 108, 112, 123, 125, 126, 128, 129, 132, 134, 136, 171, 181, 182–184
University of Austin 2
University of Milan 153
USPTO 108, 109
UWB 126

V

VDI 32, 36, 37
VeriSign 30, 117, 118
VPN 30
VTT 158

W

Weiser, Mark 28, 122
WGIG 115, 116
WLAN 24, 129
WSIS 115, 116
WTO 109, 114

Z

ZigBee technology 138

Printing: Krips bv, Meppel, The Netherlands
Binding: Stürtz, Würzburg, Germany